Field Guide to the Mangrove Trees of Africa and Madagascar

Field Guide to the Mangrove Trees of Africa and Madagascar

Henk Beentje and Salomão Bandeira

Line drawings by Juliet Williamson
Mapping by Justin Moat and Robert Frith

Kew Publishing
Royal Botanic Gardens, Kew

PLANTS PEOPLE
POSSIBILITIES

First published in 2007 by
Royal Botanic Gardens, Kew
Richmond, Surrey, TW9 3AB, UK
www.kew.org

ISBN 978 184246 135 8

British Library Cataloguing in Publication Data
A catalogue record for this book is available from the British Library

Production Editor: Michelle Payne
Typesetting and page layout: Christine Beard
Design by Media Resources, Information Services Department,
Royal Botanic Gardens, Kew

Printed and bound in the United Kingdom by Biddles

For information or to purchase all Kew titles please visit
www.kewbooks.com or email publishing@kew.org

All proceeds go to support Kew's work in saving the world's plants for life

Cover photo: *Rhizophora* mangrove swamp in Cameroon (Photo: Andrew McRobb)

Contents

Acknowledgements

We would like to thank the staff of the herbaria at BR, FT and P for access to collections, and Paul Bamps, Estrela Figueiredo, Wolfgang Kuper and Gerald Pope for interesting discussions and additional data. We are also grateful to Laurence Dorr, Andrew McRobb and Martin Cheek for providing us with some wonderful photographs.

Introduction

Mangrove trees grow on seashores, between the low and high tide marks, and along the tidal parts of rivers. They form a plant community adapted to a very changeable environment, adapted to a high degree, as plants have to cope with changing levels of water, of salt, and of oxygen. The species involved come from several plant families. In Africa and Madagascar they are distributed quite widely. Though not species-rich they form an important resource, now under threat in many parts of Africa. We hope that this publication will help with their identification and contribute to their conservation.

How to use this book

This guide concentrates on the 'true mangrove trees': those trees that have at least some specific adaptations to the ecological conditions. Such adaptations may involve special breathing roots; evergreen and thick, sometimes salt-secreting leaves; and vivipary (seeds sprouting when still on the tree). True mangrove trees should also be restricted to a tidal environment, that is, inundated by all or most high tides, and only occur in mangrove vegetation.

Other species, including such trees as *Conocarpus erectus* and *Barringtonia racemosa*, but also much smaller plants including the fern *Acrostichum aureum*, lack visible special adaptations, and occur in other habitats as well; most are rarely inundated by sea water. Such plant species are listed here as 'mangrove associate' species (see pages 40–44, 78–82). The most common associate trees occurring within mangrove vegetation, however, are included in the main species descriptions, as indicated below:

Atlantic coast (West Africa)

Avicennia germinans
Laguncularia racemosa
Nypa fruticans
Rhizophora harrisonii
Rhizophora mangle
Rhizophora racemosa

common associate: *Conocarpus erectus*

Indian Ocean coast (East Africa)

Avicennia marina
Bruguiera gymnorhiza
Ceriops tagal
Heritiera littoralis
Lumnitzera racemosa
Rhizophora mucronata
Sonneratia alba
Xylocarpus granatum

common associates: *Barringtonia asiatica*
Barringtonia racemosa
Pemphis acidula

Use the relevant key to discover the species your mangrove belongs to. Each key starts with two choices, and each choice leads you to the next part of the key, where you are given a further choice, and so on, until you find the name of your species. If you don't understand a technical term you can refer to the glossary (list of explanations), located towards the end of the book (pages 85–86). You can also identify your mangrove by looking at the list of mangrove species grouped by key characters (pages 23–25). Alternatively, if you already know the local name of your palm, you can look this up in the index at the end of the book (pages 89–90). Please note that the common names have been taken from herbarium specimens and checked against some major publications. The authors have not done an exhaustive literature search for all names.

Once you have discovered the name of your mangrove turn to the relevant entry in the species treatment section to check your plant against the detailed description. At the head of the page the scientific name is given, as is the name of the family the species belongs to and the area it occurs in (either Atlantic coast or Indian Ocean coast). The 'field characteristics' box lists some of the species' most recognisable features, and full details are given in the description. The line drawings supply a visual reference to the leaves and arrangement of flowers and fruits, and the map shows where the species occurs. Notes on the ecology, distribution, uses and conservation status of the species are also included. Finally, a list of local names appears in a box towards the bottom of the page.

A difficult habitat: how mangroves cope

Inundation by water

High tide occurs twice a day. The level of these high tides differs according to the phase of the moon: spring tides, the highest, occur at full moon and new moon; neap tides, the lowest, occur halfway in between. The highest spring tides occur when the sun is at equinox, that is when it crosses the equator and day and night are the same length. The difference between high and low tide can be considerable: for instance, in Lamu, Kenya, this difference between high and low tide may be 1 m at neap tide, and as much as 4 m at spring tides.

To cope with these variations many mangroves have to anchor themselves well, not just for this vertical difference in water levels, but also against wave action — and sometimes even because the soil in which the plant grows may be a soft sediment, such as thin mud. Stilt roots, such as those in *Rhizophora*, give a very stable support sytem. In *Avicennia*, underground (or under-mud) root systems spread far and wide — wider than the crown. Buttresses give additional stability and occur in many species, for example *Ceriops*, *Heritiera* and *Lumnitzera*.

High salt concentration

Normal salt concentration in shallow sea water is 33–39 grams per litre, but may be as high as 44 grams per litre in the Red Sea.

On the landward margin of the mangrove, sea water comes in only at the highest spring tides. The sea water then evaporates, leaving a salt concentration in the soil that is much higher than that of sea water. When the rains come, the salt is leached away and the concentration of salt is much lower than that of sea water. Plants of such areas therefore have to cope with an enormous range of salt concentration.

Mangroves need water, just like other plants. Unlike other plants, they take in mostly salt water, but their evapo-transpiration is of water vapour with a lower salt concentration. If they had no adaptations, the salt concentration in the plant would reach ever higher levels. Mangroves have special adaptations to deal with these high concentrations of salt:

• *Rhizophora mangle* actively stops salt from entering the root system, through root surface membranes that allow water to pass through but which exclude salt.
• Reduced water uptake also diminishes the amount of salt coming in, though this does not change the concentration of salt in the water, as the first method does. Mangroves with almost succulent leaves such as *Lumnitzera* have a thick leaf cuticle to reduce water loss through evaporation, and loosely packed cells to store water within the leaf. Therefore, they need less water in the first place.

- Salt-secreting leaves occur in *Avicennia* and *Laguncularia*: there are salt-secreting glands on the lower surface of the leaves, and the secretion (during the day) is much more salty than sea water; in the dry season, the salt crystallises on the lower leaf surface, and at night, when the humidity is higher, the salt dissolves and drips off.
- *Lumnitzera* and *Laguncularia* store salt crystals in the leaves, eventually shedding the leaves with the excess salt.

Resistance to fluctuation in salinity varies widely:

- high resistance: *Avicennia*, *Ceriops*
- considerable resistance: *Nypa*
- average resistance: *Rhizophora*, *Sonneratia*
- low resistance: *Bruguiera*, *Laguncularia*, *Lumnitzera*, *Pemphis*

Much salt disappears through evaporation or transpiration, but this costs water; in high rainfall zones there may be enough rain for this, but in zones with lower rainfall other sources of fresh water, such as rivers, are needed. This means that in the drier parts of Africa mangroves will be restricted to river estuaries or stream outlets.

Low oxygen levels

The plants of the mangrove zone have to cope with very low oxygen levels. At high tide, conditions are virtually anaerobic, but even at low tide the soil may be waterlogged and oxygen in the soil can be in very short supply. Adaptations to this lack of oxygen are roots above the surface, with breathing cells on the roots. Various types of roots have developed in mangroves:

- stilt roots: *Rhizophora*
- knee roots: *Bruguiera*, *Ceriops*, *Lumnitzera*
- erect, 'finger' pneumatophores: *Avicennia*, *Sonneratia*, *Laguncularia*
- main roots ribbon-like, flattened: *Xylocarpus*, *Heritiera*
- buttresses: *Xylocarpus*, *Ceriops*, *Heritiera*

All these types probably serve multiple roles in gas exchange, respiration and anchoring. They are equipped with lenticels with minute openings, through which air enters. When they are submerged, water is unable to enter these openings, as carbon dioxide is slowly trickling out.

Avicennia marina: 'finger' pneumatophores, northern Mozambique

Nypa has no breathing roots as such, but breathes through its leaves.

Most mangrove associates do not develop such specialized systems, though some palms may develop occasional pneumatophores: *Phoenix* and *Raphia* are examples.

Trouble from the start

Last but not least, the next generation of mangrove plants has to be able to survive in a hostile environment from the very start. One strategy is to have seedlings develop quite far while still growing on the parent tree; this strategy, called vivipary, is followed by *Rhizophora*, *Ceriops* and *Bruguiera*. The developed seedlings then drop and stick directly in the mud under the parent tree, and grow there, or may float away in the tide; they will develop roots as soon as they are stranded — the process of forming these roots is quite fast. The roots pull the seedling into an erect position.

Several other plant groups follow this strategy to a lesser degree, and can be called subviviparous: *Avicennia*, *Laguncularia*, *Nypa* and *Xylocarpus* have fruits in which the embryo will germinate just before or quickly after release from the parent.

Rhizophora: seedlings on parent tree

Mangroves: stilt roots, Cameroon

Pollination

There is quite a range of flower pollination agents for mangrove trees:

Avicennia	bees and other insects
Barringtonia	night-flying animals
Bruguiera gymnorhiza	insects and sunbirds
Ceriops tagal	small moths
Laguncularia racemosa	insects
Lumnitzera racemosa	small insects
Rhizophora	wind-pollinated; some bees
Sonneratia	bats, hawkmoths
Xylocarpus	mainly bees

data from Tomlinson (1986); Kalk (1995); Hogarth (1999); Bandeira (pers. obs.)

Dispersal

All mangrove fruits or seeds are dispersed by water, but the part that is dispersed differs. In *Heritiera*, *Laguncularia*, *Lumnitzera* and *Nypa* the **fruit** floats and is the unit of dispersal. In *Heritiera* there is the further adaptation that the fruit always floats with the dorsal ridge upwards, and this actually functions as a sail (Cheek & Dorr, 2007). *Laguncularia* fruit can survive floating for at least 30 days. In *Sonneratia* (and *Avicennia*) the fruit splits soon after falling, and so **both fruit and seed** act as dispersal units. *Avicennia germinans* seed can survive floating for up to four months. In *Conocarpus* and *Xylocarpus* the fruit splits and the **seed** is the unit of dispersal. In *Bruguiera*, *Ceriops* and *Rhizophora* species the seed sprouts while still on the tree, and the **seedling** is the unit of dispersal.

Xylocarpus granatum fruit: the seeds are the dispersal agents

Origin and distribution

Worldwide there are around 54 species of mangrove trees grouped in 20 genera and 16 families. According to Hogarth (1999), the mangrove habit might have evolved independently many times, with common features, such as adaptations to a saline environment, caused by convergence and not via common descent.

Species on the Atlantic coast of Africa are different from those of the Indian Ocean coast; there are no species that occur on both coasts. This is probably due to the lower temperatures of sea water in the region of the Cape where the oceans meet, as well as to the currents which generally do not allow floating fruits or seeds to pass from one ocean to another.

However, West African species of mangrove trees are also found on the Atlantic coast of the tropical Americas; and East African and Madagascan species are also found throughout the Indian Ocean and in South-east Asia.

Mangrove vegetation is distributed on shores where surf is not strong, and along tidal river estuaries. It is dependent on regular immersion in sea water, on protection from strong currents and surf, and on a sea with a minimum temperature of ± 20° C.

Mangroves on the Atlantic coast of Africa cover around 29,000 km²; the largest stands are in Nigeria, Guinea-Bissau, Guinea and Cameroon (see Table 1 on page 13).

The Indian Ocean coast has about 9,000 km² of mangroves, including those on Madagascar. The largest mangrove stands are in Madagascar and Mozambique (see Table 2 on page 14). In Madagascar mangrove vegetation is found on both the east and the west coasts, though distribution on the east coast is much more restricted due to the lower number of suitable sites.

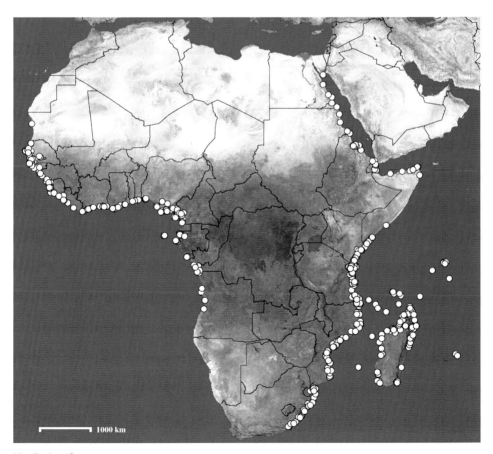

Distribution of mangrove trees

Box 1: Northernmost and southernmost populations in Africa

On the Atlantic coast, the northernmost extensive mangrove population occurs in the Parc National du Banc d'Arguin in Mauritania, at about 19°21' N (a small population occurs at 19°50'N at Ile Tidra). This is a pure stand of *Avicennia germinans*. According to Dahdouh-Guebas & Koedam (2001), in this locality *A. germinans* occurs in different forms: tall trees, wide trees, shrubs and dwarf shrubs — the last flowering profusely when as small as 30 cm high. There were no large differences in leaf morphology between these forms. Fruits and seed occur in large numbers here but only germinate successfully when protected from the Sahara wind and sun.

The northernmost mangrove on the Indian Ocean coast occurs in Egypt, where *Avicennia marina* grows in sandy soil at the edge of Sinai Desert in the Gulf of Aqaba at about 30°N and near El Suweis (Suez) at 28°N. Here the plants have numerous pneumatophores if they grow where they are regularly inundated by tide; at the landward side without tide inundation and in dry sand, few *A. marina* occur and they usually do not have pneumatophores (Hogarth, 1999).

On the Atlantic coast, the southernmost population occurs at 12°30'S near Benguela in Angola (Grandvaux Barbosa, 1970), while the estuary of Longa river, at 10°18'S between Luanda and Benguela is the southernmost extensive mangrove (Spalding *et al.*, 1997). We have records for two species, *Rhizophora mangle* and *Avicennia germinans*, for 12°30'S. Spalding *et al.* (1997) report that here *Avicennia germinans*, *Rhizophora mangle* and *R. racemosa* occur in dwarf form, the last two being less than 1 m high. This southern limit occurs in relatively low latitude due to the Benguela upwelling current which brings cold water to the West coast.

The southernmost mangrove on the Indian Ocean coast, *Avicennia marina*, occurs at the Mdumbi/Gqumbi River estuary (32°56'S) and Kobonqada River estuary (32°36'S), north of East London, South Africa. *A. marina* at the Nahoon River mouth (32°59'S), south of Mdumbi & Kobonqada, were introduced (Colloty, 2000). This occurrence so far south is due to the warm-water Mozambique current.

See Spalding *et al.* (1997) for detailed mapping based on remote sensing.

Table 1: Atlantic coast species by area

(North to south)	Area (km²)	Main mangrove areas	Avicennia germinans	Laguncularia racemosa	Nypa fruticans	Rhizophora harrisonii	Rhizophora mangle	Rhizophora racemosa
Morocco		(No mangrove)						
Western Sahara		(No mangrove)						
Mauritania	1	Senegal River	✓					✓
Cabo Verde		(No main mangrove area)						
Senegal	1,830	Saloum River, Casamance River	✓					
Gambia	747	Gambia River	✓	✓		✓	✓	✓
Guinea Bissau	3,649	North coast	✓	✓		✓	✓	✓
Guinea	3,083	All along the coast	✓	✓		✓	✓	✓
Sierra Leone	1,695	North coast	✓	✓		✓	✓	✓
Liberia	427	Lake Piso, central coast	✓	✓		✓	✓	✓
Côte d'Ivoire	644	Assini to Cavally River	✓	✓				✓
Ghana	214	West coast	✓	✓		✓	✓	✓
Togo	26?	Togo Lagoon	✓			✓		✓
Benin	17	Benin Lagoon	✓					
Nigeria	11,134	Niger Delta, Cross River Estuary	✓	✓	✓	✓	✓	✓
Cameroon	2,494	Coast east and west of Mt Cameroon	✓	✓	✓	✓	✓	✓
São Tomé & Príncipe	1	Near Porto Alegre	✓	✓		✓		
Gabon	1,759	Como R, Ogooué River	✓	✓		✓	✓	✓
Equatorial Guinea	277	Mbini, Muni and Ntem River	✓	✓				✓
Congo (B)	188	Malonda & Conkouati	✓	✓			✓	✓
Congo (K)	374	Congo River	✓	✓		✓	✓	✓
Angola	607	Cabinda	✓					✓
Namibia		(No mangrove)						
South Africa		(No mangrove)						

Source for area coverage: Spalding et al. (1997)

Table 2: Indian Ocean coast species by area

(South to north)	Main mangrove areas	Area (km²)	Avicennia marina	Bruguiera gymnorhiza	Ceriops tagal	Heritera littoralis	Lumnitzera racemosa	Rhizophora mucronata	Sonneratia alba	Xylocarpus granatum
Africa										
South Africa	St Lucia	334	✓	✓	✓		✓	✓		
Mozambique	Zambezi Delta area	3,961	✓	✓	✓		✓	✓	✓	✓
Tanzania	Rufiji Delta	2,456	✓	✓	✓	✓	✓	✓	✓	✓
Kenya	Lamu, Mida	961	✓	✓	✓	✓	✓	✓	✓	✓
Somalia	Juba/Shebele Estuary	910	✓	✓	✓	Drift		✓		
Djibouti	(no main mangrove area)	10	✓	✓				✓		
Eritrea	Mitsiwa Bay	581	✓	✓	✓		✓	✓		
Sudan	(no main mangrove area)	937*	✓	✓				✓		
Egypt	(no main mangrove area)	861*	✓					✓		
Indian Ocean Islands										
Madagascar	North-west coast	3,403	✓	✓	✓	✓	✓	✓	✓	✓
Seychelles	Aldabra	29	✓	✓	✓	✓	✓	✓	✓	✓
Comoros	(no main mangrove area)	26	✓	✓	✓		✓	✓		
Reunion	(no main mangrove area)	10		✓				✓		
Mayotte	(no main mangrove area)	10								
Mauritius & Rodrigues	(no main mangrove area)	1-2		✓				✓		

Main source for area coverage: Spalding et al. (1997); other sources: Kokwaro (1985), Roger & Andrianasolo (2003), Semesi (1998), Taylor et al. (2003); figures for Mozambique from Bandeira. *Area figures for Sudan and Egypt are probably overestimates (Spalding et al. 1997).

Zonation

Mangrove vegetation varies from single-species stands, for example in the northernmost and southernmost parts of the Red Sea, Mauritania and central Angola; to rich and diverse communities, such as those of the Niger or Zambesi Deltas. In the more diverse communities there is often a distinct zonation of species.

Within their distribution area, occurrence of the species varies considerably. In some sites the vegetation type may be rich with high species diversity (high for mangrove vegetation, that is), while a few kilometres away a few straggly trees may be struggling to stay alive. Dwarf and tall mangroves of the same species may also occur within a relatively small area, the first occurring in flat areas with high salinity, little fresh water flow and usually as the only species, and the second along river banks where fresh water lowers the salt concentration of the sea.

The most diverse and most complicated zonation happens on shallow slopes by river estuaries. Along rivers there is less scope for zonation, and usually only few species occur. Zonation seems mostly dependent on salinity tolerance and substrate/soil type.

Atlantic coast

In the richer mangrove areas on the Atlantic coast of Africa, *Rhizophora racemosa* is a pioneer on mud. Once this species has become established and a root mat has formed, the habitat accumulates more mud enabling *Rhizophora harrisonii* and *R. mangle* to grow. Where the soil level is raised to a level where it is not inundated so regularly, the salinity increases (through evaporation) and the habitat becomes too extreme for *Rhizophora*; *Avicennia* can withstand this higher salinity and takes over. The zone between the normal high tide mark and the spring tide mark is flooded twice a month. *Laguncularia* is the pioneer in this habitat and is later replaced by *Avicennia*.

Zonation will often be as follows: on the seaward side there is a belt of *Rhizophora* with *R. racemosa* the pioneer on the outside, *R. harrisonii* in the middle, and *R. mangle* on the inside. This is followed by a belt of *Avicennia*, and finally a belt of *Laguncularia*. *Conocarpus erectus* and *Dodonaea viscosa* can dominate the inward margin with more inland vegetation types, where tides reach only rarely. This same zonation, or succession, occurs on the tropical Atlantic coasts in the Americas, with the exception of *Laguncularia*.

Sonneratia is a pioneer on more open coasts; although in areas where the seaward substrate is firm and sand, rather than mud, is deposited, *Avicennia* is the pioneer. When *Avicennia* establishes itself in such situations, it can sometimes cause mud desposits, which in turn leads to *Rhizophora* taking over.

Along the River Gambia mangroves extend 160–190 km upstream. *Rhizophora* forms the seaward belt, restricted to the zone flooded daily; *Avicennia* forms an open bushland up to and at the limit of spring tides. Although the dry season may last for as long as eight consecutive months, the supply of river water keeps water salinity upstream as low as 5 grams per litre (as opposed to 35 grams per litre at the river mouth) enabling mangroves to grow over 20 m high.

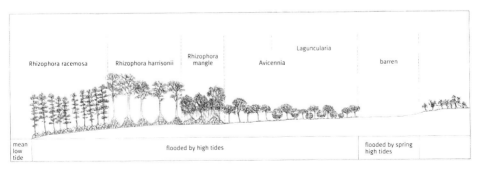

Fig. 1. Common zonation in rich West African mangroves

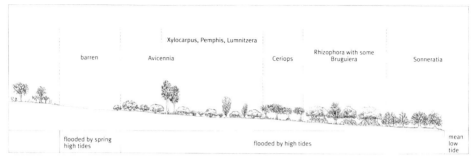

Fig. 2. Common zonation in rich East African mangroves

Indian Ocean coast

In the most diverse mangrove areas of the Indian Ocean coast, *Avicennia* and *Sonneratia* are generally the pioneers, though usually only one of these forms the seaward zone: *Avicennia* seems to prefer sand, while *Sonneratia* prefers fine silt, mud and firm substrates. *Rhizophora* is dependent on mud, and will not tolerate salinity higher than sea water, therefore it often forms stands at river mouths. *Bruguiera* is most tolerant of fresh water and occurs where the water table is high; it can be very common on rain-rich coasts. *Ceriops* is the main species of drier areas; while *Lumnitzera*, *Pemphis* and *Xylocarpus* are the most landward, though stunted *Avicennia* can occur where there is a zone of very high salinity on rarely inundated areas.

Walter & Steiner (1936) describe zonation in the Tanga area of Tanzania. They stress that zonation is not exact and in neat rows, there are many irregularities, depending on water depth, water salinity and texture of alluvium. Similar patterns are described by Kairo *et al.* (2002) for Mida Creek in Kenya, and roughly similar ones by Macnae & Kalk (1962) for Inhaca Island, Mozambique, White (1983) for Gazi Creek, Kenya and by Barbosa *et al.* (2001) for Mozambique.

A typical zonation pattern is formed by pure stands of *Sonneratia alba* in deepest water, followed by a belt of *Rhizophora mucronata* where there is slightly higher salt concentration. Sometimes a narrow belt of *Ceriops tagal* is present, also along small channels far into the next zone. In brackish water at river mouths *Rhizophora* is often the commonest species. Along creeks there is a fringe of *Rhizophora* and *Xylocarpus*. *Bruguiera gymnorhiza* occurs scattered throughout the *Rhizophora*, *Ceriops* and *Avicennia* zones. The most landward zone, with the shortest inundation period, is formed by bushy *Avicennia marina*, together with *Lumnitzera racemosa*, *Heritiera littoralis*, *Xylocarpus granatum* and *X. moluccensis*. In this *Avicennia* zone salt concentration is similar to that of the sea during high tide, but can be much higher during low tide. Small hummocks can have associated species such as *Arthrocnemon* and *Sporobolus virginicus*. Behind the *Avicennia* zone, on the land side, the salt content of the soil varies so greatly that often vegetation is almost absent; this zone is flooded only twice a year. In Mozambique and South Africa *Avicennia* replaces *Sonneratia* as the most seaward belt-forming species.

Uses of mangroves

'Mangroves are an entrepreneurs' dream . . . they produce raw material (lignocellulose) from sea water by using renewable energy sources (sunlight, and tidal energy bringing in fresh nutrients).' (Tomlinson, 1986).

Mangrove vegetation is an important resource, as it provides a nutrient-rich environment for young fish of many species, as well as a rich habitat for marine invertebrates such as prawns. Semesi (1998) remarks that in eastern Africa the richest mangrove areas (Rufiji Delta and Sofala) are close to the richest shrimp and prawn areas just offshore, and suggests that the mangroves form the nurseries for these. Small-scale fishing by local communities is certainly directly linked to inshore ecosystems like mangroves.

Rhizophora mucronata, northern Mozambique

Mangrove vegetation also stabilizes the shoreline, preventing erosion, and has proved to be an important coastal defense against cyclones and tsunamis. Where mangroves have been destroyed, such natural disasters cause far more damage. By catching the sediment coming down rivers they promote, to a certain extent, land build-up in tidal areas. Mangrove vegetation also provides a natural effluent/sewage treatment, as nitrates and phosphates are absorbed, mainly by microflora.

Mangrove vegetation is traditionally harvested by local communities for tannins (used to preserve fishing nets, ropes and sails), fuelwood, and minor forest products such as medicines and honey. The leaves of *Avicennia marina* are used for animal fodder. As the wood of many species is hard and water-resistant, it is used extensively for construction and boat-building materials, and it is used in carpentry and crafts to produce tool handles, telegraph poles, fish traps and railway sleepers. Several taxa such as *Avicennia*, *Laguncularia* and *Sonneratia*, but not *Rhizophora*, will re-sprout from cut stumps and are therefore a renewable resource. Poles of these species are used in house walls and roofs, as main rafters and corner poles; mangrove poles are widely used in coastal building construction.

Mangrove areas in Africa have also been converted into shrimp pond and salt pan areas. Although aquaculture activities in Africa in general are modest, mangroves and adjacent areas can be used for farming, for example, finfish or oysters. But this is an activity to be undertaken only if resource sustainability is guaranteed and the vital role of mangroves in the ecosystem is maintained.

Mangrove habitats represent a diverse living resource that is very valuable for the economy, both on a local and regional scale. Reducing mangrove vegetation impacts upon fish stocks by diminishing the number of nurseries for young fish, a relationship yet to be quantified.

To sum up: mangroves are a renewable resource if treated carefully, and are valuable for plant products, fishery resources, water supply resources and on a small scale for agricultural and forage resources. They are important in shoreline protection, and help with sediment regulation and thereby water quality maintenance.

Threats and conservation

Mangrove vegetation is decreasing in many parts of Africa, particularly around urban areas and large villages. This is mainly due to increasing human population and increasing poverty, leading to more and more tree cutting for fuelwood, building material and urban development. Large mangrove areas are also converted into salt pans or to agricultural land, and to a lesser extent for aquaculture purposes (oyster, shrimp and fish ponds). In Guinea, 1,400 km² (over 40 per cent) of existing mangrove swamp has been converted to rice fields; by 1993, 620 km² of this area had been abandoned and is now almost barren (Spalding *et al.*, 1997). Coastal developments, mainly for infrastructure and tourism, further threaten mangrove areas; and the growing population has led to increased pollution from industry and agriculture (for example from crude oil, heavy metals and pesticides).

Many coastal towns in Africa, such as Freetown, Abidjan, Lagos, Durban, Beira and Dar es Salaam, have developed at the expense of mangrove forests. Around 14,000 ha of mangrove forests vanished in Mozambique due to the combined effects of fuelwood harvesting, salt pan production and town development in the last 30 years (Barbosa *et al.*, 2001). In Kenya, around 70 per cent of the people living on the coast rely on mangrove poles for construction (Kairo *et al.*, 2001). Dam construction upstream may reduce the amount of fresh water available to the mangroves; this is one of the reasons for mangrove drying and delta erosion in the Zambezi Delta in Mozambique. The construction of canals in lagoons has similarly destroyed mangroves in Côte d'Ivoire. In Madagascar firewood is being cut in the mangroves for fish and prawn processing (Semesi, 1998). Over the past decade some salt flats and mangrove areas have been transformed to shrimp farms (Roger & Andrianasolo, 2003). For detailed country assessments see Spalding *et al.* (1997).

The most pressing problem is tree felling for fuelwood and building material: if mangrove vegetation is to be conserved it is vital this over-exploitation ceases. Oil prospection has also led to the degeneration of mangrove vegetation. Replanting will help rehabilitate the habitat, but it is just as important that the mangroves are harvested sustainably, allowing time for natural regrowth.

Mangrove vegetation, northern Mozambique

Mangrove creek in Kwazulu Natal, South Africa

Mangrove area near Mombasa, Kenya

Possibly this could be achieved by selective harvesting and adopting a rotational system of cutting. Alternative sources of fuel, for example natural gas, should also be exploited.

Deforestation is also caused by urban development. Any extensive mangrove vegetation in urban or developing areas should be left intact wherever possible. This may be achieveable through community involvement with sustainable harvesting and ecotourism, which would help boost local incomes. Tourism is generally perceived as a threat, but carefully conceived ecotourism could provide some economic benefits without endangering the mangrove habitat. Possibilities include the construction of pathways above the high tide mark, enabling tourists to enjoy bird watching or the observation of crabs and mud skippers; mangrove nature 'walks' by kayak; ecology tours and environmental education classes.

Changes in the land use of mangrove areas can be detrimental to the mangroves. Dam construction results in irregular fresh water supplies reaching the coast, while salt pan construction and salt extraction alters the salinity level of the water. The problems associated with dam construction could be alleviated by the relevant authorities carefully considering whether or not additional dams are necessary, and then balancing the perceived advantages against the disadvantages dams bring to fisheries. Where dams are necessary, the dam authorities could help the

Rhizophora mucronata and seedlings, Luchete, northern Mozambique

vegetation and marine life by releasing more water during the rainy season; thereby imitating natural flood regimes. The authorities should also try to avoid building salt pans near urban areas in order to minimize the contamination of well water with saline water. Surrounding mangrove vegetation could be protected from increased levels of saline water if the authorities ensured adequate drainage of pumped salt water.

Mangroves can also suffer when agriculture leads to fresh water being drained from rivers with mangrove forest. This problem could be eased by modernising agricultural production methods in order to save more river water, and by allowing some drainage so that fresh water returns to the river.

The development of aquaculture is the final cause for concern. Aquaculture destroys natural nurseries for marine fish and crustaceans, while the resulting dip in fish stocks damages fishermen's livelihood. Furthermore, clearing mangrove areas impacts on coastal protection against natural disasters such as cyclones and tsunamis.

Mangrove with *Xylocarpus granatum*, near Mombasa, Kenya

Mangrove rehabilitation

Replanting

Always involve the local community: success of large-scale mangrove tree planting depends on local community involvement and understanding.

1. Choose your replanting sites by identifying areas where trees have been heavily cut, where the need for mangrove wood is high, or where there is risk of marine floods or coastal erosion.

2. Start your nursery by collecting the viviparous fruits in the wild and planting them in a place where you can check them on a regular basis for at least the first three months.

3. Plant the new plants in places where similar species are likely to occur naturally. The choice of replanting site is critical. Conditions such as salinity and soil type have to be taken into account: compare this with natural populations.

4. Monitor your replanting site for the first few months and keep it free from other plants such as *Acrostichum aureum* and *Nypa fruticans*.

Mangroves on abandoned salt pans or shrimp farms can be restored by removing the dikes, thereby restoring normal water flow and allowing natural regeneration of seedlings.

Acid soils in shrimp aquaculture ponds are also a major problem. This is due to oxygenation of pyrites (metal complex in solid form; occurring normally in anoxic mangrove soils) which gives rise to the formation of sulfuric acid. In order to rehabilitate these acid soils add lime and fresh water.

New areas (products of sand accretion) can be reforested by planting mangroves — while avoiding the destruction of other vegetation such as sea-grass beds!

Keys

Note: some of the more common associates are included as they occur together with true and obligatory mangrove species.

Key to Atlantic coast species

1. Leaves compound; palm tree .... *Nypa fruticans* (p. 32)
   Leaves simple 2
2. Leaves opposite 3
   Leaves alternate .... *Conocarpus erectus* (p. 28)
3. Stilt roots absent; stipules absent 4
   Stilt roots present; stipules present 5
4. Petiole without glands near the top; breathing roots pencil-
      like, many .... *Avicennia germinans* (p. 26)
   Petiole with 2 glands near top; breathing roots peg-like,
      few above ground and short-lived .... *Laguncularia racemosa* (p. 30)
5. Fresh leaf midrib straw-coloured; inflorescence with 2–4
      flowers, with a stalk 3–4 cm long; flower bud angular or
      semi-twisted .... *Rhizophora mangle* (p. 36)
   Fresh leaf midrib never straw-coloured, usually red-
      coloured; inflorescence many-flowered and much
      branched, with a stalk more than 5 cm long; flower
      buds pointed or blunt 6
6. Flower buds blunt; flower stalk 3–4 mm long, stout; sepals
      greenish .... *Rhizophora racemosa* (p. 38)
   Flower buds pointed; flower stalk 6–15 mm long, slender;
      sepals greenish-yellow .... *Rhizophora harrisonii* (p. 34)

Key to Indian Ocean coast species

1. Leaves simple 2
   Leaves compound; fruit like a cannonball, 14–25 cm
      across .... *Xylocarpus granatum* (p. 72)
2. Leaves alternate 3
   Leaves opposite 6
3. Leaf with silvery scales beneath; fruit woody and in
      clusters, boat-shaped, ridged, 6–8 cm long .... *Heritiera littoralis* (p. 58)
   Leaf without scales beneath 4
4. Leaves 5–50 × 2–20 cm; breathing roots sometimes
      present as knees or loops 5
   Leaves 2–8 × 1–3 cm; stem with angular buttresses; knee
      roots present; fruit small, less than 1.2 cm long .... *Lumnitzera racemosa* (p. 60)
5. Leaf margin entire; flowers fewer than 10 per head; fruit
      cubical, 8–11 cm .... *Barringtonia asiatica* (p. 48)
   Leaf margin usually toothed; flowers many; fruit longer
      than wide, 3–9 × 2–5 cm .... *Barringtonia racemosa* (p. 50)
6. Leaf silvery-green hairy beneath; breathing roots pencil-
      like .... *Avicennia marina* (p. 46)
   Leaf hairless beneath (may have corky warts) 7
7. Stilt or knee roots present; stipules present 8
   Finger roots present or absent; stipules absent 10
8. Knee roots present; leaf smooth, hairless, 1–5 cm wide 9
   Stilt roots present; leaf cork-dotted beneath, 4–9 cm wide ...... *Rhizophora mucronata* (p. 64)
9. Leaves dark green, with subacuminate to acute tip;
      inflorescence solitary flowers; calyx 2–3.5 cm long; fruit
      top-shaped, to 4.5 cm long .... *Bruguiera gymnorhiza* (p. 52)

Leaves yellow-green, with rounded tip; inflorescence of
4–8 flowers; calyx ± 0.5 cm long; fruit egg-shaped,
1.5–2.5 cm long . . . . *Ceriops tagal* (p. 56)
10. Conical breathing roots present; fruits large, 2–3 x 3–4
cm, top-shaped or apple-like . . . . *Sonneratia alba* (p. 68)
Breathing roots absent; fruit small, ± 5 mm, spherical,
included in calyx . . . . *Pemphis acidula* (p. 62)

Mangrove species grouped by key characteristic

Root type/trunk base

stilt roots	*Rhizophora*, all species
knee roots	*Bruguiera, Ceriops, Lumnitzera*
main roots ribbon-like, flattened	*Xylocarpus, Heritiera*
erect pneumatophores	*Avicennia, Sonneratia, Laguncularia*
buttresses	*Xylocarpus, Ceriops, Heritiera*
no visible roots above ground	*Barringtonia, Heritiera, Nypa*

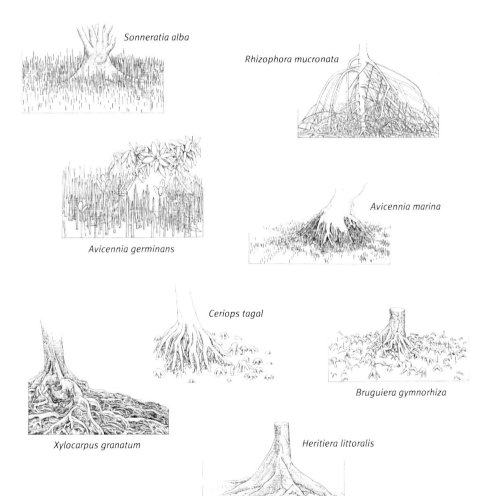

Sonneratia alba

Rhizophora mucronata

Avicennia marina

Avicennia germinans

Ceriops tagal

Bruguiera gymnorhiza

Xylocarpus granatum

Heritiera littoralis

Flowers

Flowers showy, over 1 cm long
Flowers less than 1 cm long

Barringtonia, Bruguiera, Sonneratia
other species

Fruit

Avicennia species
Barringtonia asiatica
Barringtonia racemosa
Bruguiera gymnorhiza
Ceriops tagal

2–3 cm long, short beak
top-shaped, 8–12 cm
ovoid and 4-angled, 2–4 cm long
top-shaped, to 4.5 cm long (sprouting on tree)
pear-shaped, 1.5–2.5 cm long (sprouting on tree)

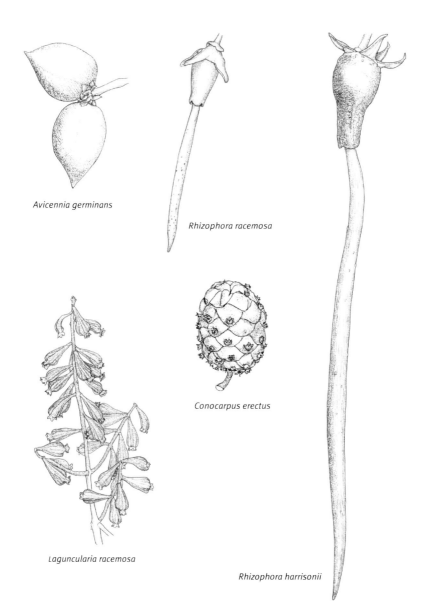

Avicennia germinans

Rhizophora racemosa

Conocarpus erectus

Laguncularia racemosa

Rhizophora harrisonii

Heritiera littoralis	ellipsoid, 6–8 cm long, ridged
Laguncularia racemosa	club-shaped, 1.5–2 cm long
Lumnitzera racemosa	tapered to both ends, 1–1.2 cm long
Nypa fruticans	big round aggregate heads
Rhizophora species	pear-shaped, 2–3 cm long (sprouting on tree)
Sonneratia alba	pear-shaped, 3–4 cm long
Xylocarpus granatum	subglobose, 14–25 cm across

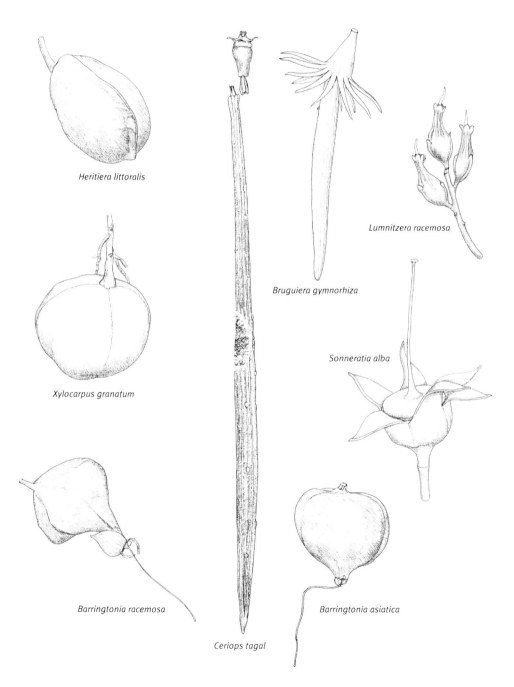

Heritiera littoralis

Lumnitzera racemosa

Bruguiera gymnorhiza

Xylocarpus granatum

Sonneratia alba

Barringtonia racemosa

Ceriops tagal

Barringtonia asiatica

Avicennia germinans (L.) Stearn

(= *Avicennia africana* P. Beauv.; *A. nitida* of some authors)

Avicennia germinans is the most widespread mangrove on the Atlantic coast of Africa, growing as a pioneer on sand, as well as in very saline conditions.

Field characteristics

The combination of opposite leaves and the abundance of simple finger roots, coupled to the absence of stilt roots.

Description

Small to medium-sized evergreen tree to 20 m tall, but may flower and fruit when as short as 1 m in extreme situations. Crown dense and usually rounded. **Bark** smooth and grey or brown on young trunks, becoming thick, dark brown, scaly and fissured when older. **Breathing roots** pencil-like pneumatophores up to 30 cm high from subterranean roots cover the surface surrounding the trunk. **Young twigs** with swollen nodes, finely hairy when young. **Leaves** opposite, leathery, elliptic, 3–18 × 1–8 cm, hairless, shiny above, dull olive-green and grey-scurfy with salt crystals beneath. **Flowers** held in crowded round heads in leaf axils or at the ends of shoot, but the central stalk lengthens when the lower flowers drop off; white with a yellow centre, about 6 mm long; calyx with 5-lobes. Corolla with a tube and 4 lobes. Stamens 4. **Fruit** an obliquely ellipsoid green capsule 2–5 × 1–2 cm, flattened laterally, with a short beak, velvety-hairy, opening by two valves; usually a single black seed develops, in which the embryo is already sprouting before the fruit drops. The fruit and seed can survive floating for up to four months.

Ecology and distribution

This species withstands the most salty conditions of all West African mangroves. It therefore grows in extreme situations; it is also a pioneer on sandy deposits. On mud, this species only appears when *Rhizophora* has become established; then it can form a zone behind the *Rhizophora*, between the high water mark and the spring tide mark.

Flowering times: Sierra Leone Dec–Jun; Ghana Apr, Jun–Jul, Oct; Nigeria Apr–Jun, Aug–Dec.

Fruiting times: Nigeria Aug–Dec.

Occurs from Mauritania to Angola and is the most widespread West African mangrove. It also occurs in tropical America, on both the Atlantic and Pacific coasts.

Uses

This species will re-sprout from cut stumps, and so can be a renewable resource. The bark is used in tanning (up to 12 per cent tannin) and produces a red dye. The wood is used for construction timber. It is often used for firewood and makes good charcoal. Bark powder decoctions are used widely against skin complaints. Salt is prepared from the leaves. The seed is eaten in famine times, but may be poisonous when raw or cooked improperly. This is an important honey plant, from which bees produce white honey of high quality.

Conservation status

Widespread, least concern (LC) on a global basis.

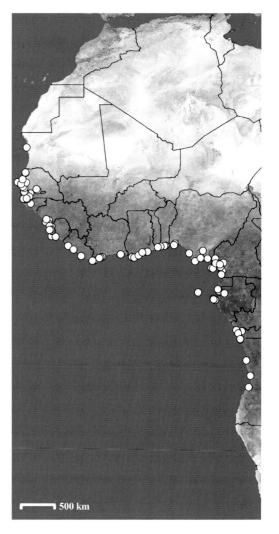

500 km

Common names

Black mangrove; Jaia-guwi, Gbeleti (Mende, Sierra Leone), Bue, Bue-Dinte (Sherbro, Sierra Leone), Ka bure (Temne, Sierra Leone), Guitouconon, Miarco (Mandingue, Senegal), Mbougari (Wolof, Senegal), Sanaj (Peulh, Senegal), Saanar (Gambia mouth, Senegal), Skasut (Sierra Leone), Bandjo (Mabula, Cameroon), Aguirigui (unspecified, Gabon), Iguiri (Port Gentil, Gabon), Nvandi (Banana, Congo Kinshasa).

Conocarpus erectus L.

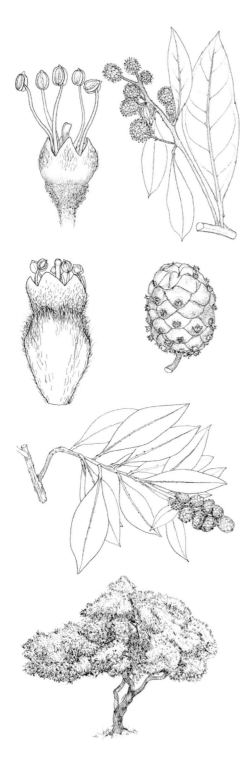

Conocarpus erectus is not a true mangrove, as special adaptations seem to be lacking: there are no stilt roots or breathing roots. This species also occurs in more inland vegetation types. It is included here as it commonly occurs in mangrove vegetation.

Field characteristics
The slightly winged branches are unmistakable, as is the leaf stalk with two glands.

Description
A shrub 2–4 m tall, sometimes growing into a tree to 20 m tall. Crown spreading. Trunk to 30 cm in diameter. **Bark** grey or brown, with age becoming thick, rough and fissured. **Breathing roots** none. **Young twigs** angular to slightly winged, yellow-green, usually hairless. **Leaves** alternate, leathery-succulent, lanceolate, 4–10 × 1.5–3.5 cm, tip pointed, rather glossy and medium green above, on the lower surface with small pockets in vein axils, hairless; veins in 6–8 pairs, curved; leaf stalk winged, with two glands. **Flowers** in dense ellipsoid to almost round pale green heads 1–1.5 cm long, these heads arranged in clusters; heads are male on some trees, female on others; flowers sessile, small (less than 3 mm long). Calyx with 5–6 lobes. Petals absent. Stamens 5–7, yellow; disk orange; style curved. **Fruit** of little trapezoid achenes, 5 mm long and 2 mm wide, topped by calyx remnants, hairless, brown, grouped in cone-like heads, 15 × 11 mm, brown-red. Seed curved, 3 × 1 mm, not sprouting on the tree.

Combretaceae

Ecology and distribution

Grows on sea shores and sandy lagoon shores as well as in mangrove vegetation; at the landward edge of mangrove and grading into normal, non-seawater-adapted vegetation.

Flowering times: not enough data.

Occurs from Senegal to Congo (Kinshasa), and in tropical and subtropical America.

Uses

The bark is sometimes used in tanning. The wood is heavy, hard and durable, and fine-textured; it is used in boat construction and is highly valued for firewood and charcoal. Cut branches are reported to grow readily when stuck in the ground.

Conservation status

Widespread, least concern (LC) on a global basis.

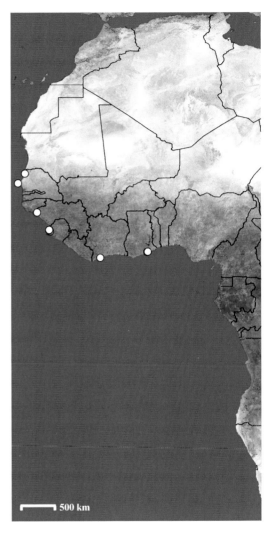

500 km

Common names:

None found for African specimens; in the Americas known as 'buttonwood'.

Laguncularia racemosa Gaertn.

Laguncularia racemosa is a shrub or small tree of the landward side of the mangrove, with peg roots hidden underneath the surface.

Field characteristics

The red leaf stalk with its glands is unique to this species.

Description

Shrub or small tree, to 20 m tall but more usually 2–4 m tall, often with multiple trunks. Crown rounded or irregular. Trunk up to 30 cm in diameter. **Bark** grey-brown, smooth, later becoming grooved, then rough and fissured. **Breathing roots** erect peg roots present, but underground, and only short-living roots above the surface; has been described as having pneumatophores 'facultatively' (Jenik, 1970). **Young twigs** greenish- to reddish-brown, hairless. **Leaves** opposite, leathery, fleshy, olive-green, obovate or elliptic, 5–11 × 3–6 cm, hairless; base rounded, tip rounded with a small point, sometimes notched; lower surface with crater-shaped glands near the margin and on the veins, secreting salt; leaf stalk red, with a pair of glands near top, which produce a sweet nectar. **Flowers** sessile in axillary and terminal spikes, bell-shaped, whitish. Calyx tube 4–6 mm long, increasing in fruit. Petals tiny (± 1 mm), soon falling. Stamens 10, included. **Fruit** sessile, club-shaped and topped by the persistent calyx, grey-yellow, 1.5–2 × 0.8–1 cm, flattened, ribbed longitudinally, velvety, leathery, indehiscent; one-seeded. Seed 15 × 5 mm, developing while the fruit is still on the tree or floating.

Combretaceae

Ecology and distribution

A plant of the landward side of mangrove zone, between the high tide mark and the spring tide mark. Possibly it is a pioneer in this habitat.

In seasonal climates this tree is dioecious, meaning there are male and female trees; but trees with flowers of both sexes also occur. Fruiting is abundant. Fruits can survive floating for up to thirty days.

Flowering times: Nigeria Apr–Jun, Dec.

Fruiting times: Nigeria Apr–Jun, Dec.

Occurs from Senegal to Angola; also grows on the Atlantic and Pacific coasts of tropical America.

Uses

Browsed by camels. The bark is used for tanning (Mpengwe, Gabon). The wood is hard, heavy, dense and strong, and is used for firewood or charcoal. A rapid grower, which can flower and fruit when two years old. This species will re-sprout from cut stumps. A honey tree.

Conservation status

Widespread, least concern (LC) on a global basis.

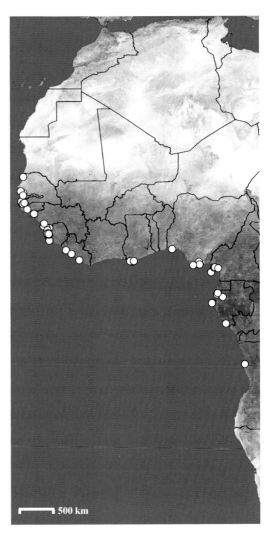

Common names:

White mangrove; Diain mangui (Wolof, Senegal: 'slave of the mangrove'), Bak (Seren, Senegal), Chemchem-de (Sherbro, Sierra Leone), Tarafe preto, Oellha (Guinea-Bissau), Orke (Brass, Nigeria), Maganga (Malimba, Cameroon), Ambianbiolan (Gabon, unspecified), Ntehigizigi (Mpengwe, Gabon).

Nypa fruticans Wurmb.

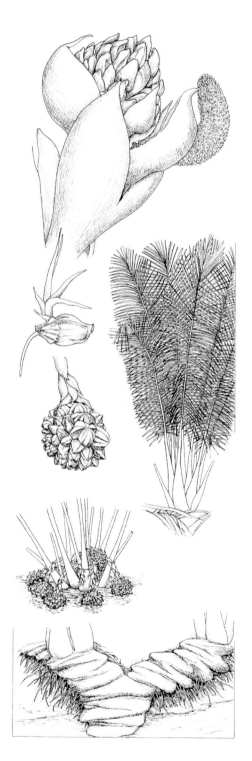

Nypa has been imported from Asia and is establishing itself agressively, crowding out native plants.

Field characteristics

The only palm growing in true mangrove vegetation, though several other palm species (all with trunks) may be found on the landward side of the mangrove zone.

Description

A small palm with an underground branched 'trunk' to 70 cm across; the bases of the leaves spring from this in a dense cluster and these bases are immersed in mud or are underwater; underground parts look like cowpats. **Bark** is underground. **Breathing roots** absent. **Young twigs** not visible. **Leaves** in a crown, to 7 m long, the leaf stalk 1–2 m long, the leaf itself pinnate with many (30–40) leaflets up to 70 cm long on each side of the leaf midrib. **Flowers** many, grouped in a large long-stalked inflorescence surrounded by large bracts (modified leaves); female flowers (and later fruits) in a congested round head at the top, male flowers in dense club-shaped spikes on side branches; flowers with 6 petals up to 5 mm long. **Fruit** in a large round head, with many aggregated fibrous fruits; fruits with the seed embryo already protruding before the seed drops.

Ecology and distribution

This species grows in sheltered tidal estuaries, forming pure stands.

Spreads in colonies by branching and possibly dispersing by broken-off roots. *Nypa* replaces mangroves where the mangroves are in poor condition (for example through felling for charcoal, or around petrochemical industry sites); this replacement is possibly negatively affecting fish breeding.

Flowering times: not enough information.

Introduced from Singapore in 1906 and 1912 and again in 1946 and now spread from Lagos to the Wouri Estuary near Douala (Sunderland & Morakinyo, 2002). Its natural distribution is from India to Queensland.

Uses

In Asia it is used for thatch and palm wine, and the immature fruits are eaten; in Africa it is sometimes sometimes used for thatching.

Conservation status

Widespread, least concern (LC) on a global basis.

500 km

Common names:

No local names, but often referred to as 'nipa'.

Rhizophora harrisonii Leechman

Rhizophora harrisonii is a common and sometimes dominant mangrove tree in the Niger Delta, growing inside the *Rhizophora racemosa* belt. *R. harrisonii* is probably a hybrid between *R. racemosa* and *R. mangle*, and sets little ripe fruit.

Field characteristics

All three *Rhizophora* have distinctive stilt roots; this species differs from its nearest relative, *Rhizophora racemosa*, by the longer, more slender and less compact inflorescence; and by the fruit, which is similar but smaller, and in which the root develops to less than 20 cm before the fruit drops.

Description

Shrub or tree to 12 m. **Bark** smooth, becoming furrowed with age. **Breathing roots** many stilt roots from lower part of trunk. **Young twigs** swollen at nodes. **Leaves** opposite, with stipules between the leaf stalks, narrowly elliptic, 7–13 × 2.5–5 cm; midrib often reddish; underside with fine black cork dots, visible when dry. **Flowers** many (usually 8–32), in long (7–12 cm) axillary clusters; buds pointed, greenish yellow. Sepals 12 mm. Corolla with 4 petals, deciduous. **Fruit** bottle-shaped, about 3.5 × 1.8 cm, germinating on the tree with the root reaching 20 cm long while still on the tree.

Ecology and distribution

In mangrove vegetation inundated at all high tides. Can be dominant in the middle belt, for example in the Niger Delta, forming dense thickets with many trunks.

Flowering times: Nigeria FEB–MAR, JUN–DEC.

Fruiting times: Nigeria in most seasons.

Occurs from Senegal to Gabon (?Angola); Atlantic coast of tropical America.

Uses

The bark used for red dye and tanning, and the wood for firewood and construction.

Conservation status

Widespread, least concern (LC) on a global basis.

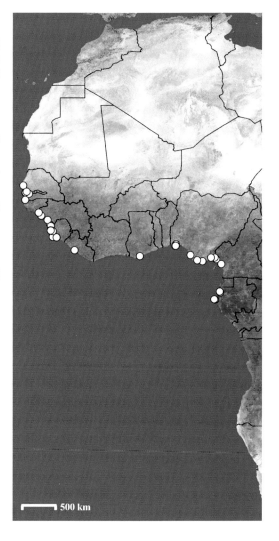

500 km

Common names:

Manko (Mandinka, Gambia), Dengi (Mende, Sierra Leone), Jaiei or Jaia-Lelei (Mende, Sierra Leone), Kinsi (Susu, Sierra Leone); Sule or Shule (Sherbro, Sierra Leone), Shundinte-le (Sherbro, Sierra Leone), Suthe-le (Sherbro, Sierra Leone).

Rhizophora mangle L.

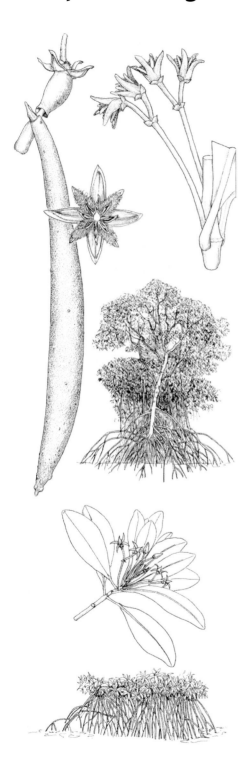

Rhizophora mangle is the shrubbiest of the three West African *Rhizophora*, and prefers drier situations than the other two.

Field characteristics

All three *Rhizophora* have distinctive stilt roots; this species differs from the other *Rhizophora* by the few flowers in the inflorescence.

Description

Shrub or tree to 5 m tall, though in tropical America it may grow to a large (to 18 m) tree. Trunk to 20 cm in diameter. **Bark** thin and smooth and grey or grey-brown, with age becoming thick and furrowed. **Breathing roots** stilt roots abundant from the lower trunk. **Young twigs** swollen at the nodes. **Leaves** opposite, with stipules 6–8 cm long, shiny green, narrowly elliptic, 5–15 × 2.5–5 cm, edges slightly recurved; tip blunt; underside with abundant fine black cork dots when dry. **Flowers** 2–5 per axillary cluster, scented; buds angular. Sepals greenish yellow, midrib straw-coloured. Corolla with 4 whitish petals 7–8 mm long, hairy inside, deciduous. Stamens 8–12, inserted on a disc. **Fruit** olive green with a brown tip to dark brown, indehiscent, 2–3.5 cm long, germinating on the tree with the root reaching 20 cm long while still on the tree. Fruit and seedling float.

Ecology and distribution

Grows on mud established by *Rhizophora racemosa*, and on the landward edge of the *Rhizophora* zone, inside the *Rhizophora racemosa* and *R. harrisonii* belts. Usually on the landward side, on firmer soil.

Flowering times: not enough data for Africa; throughout the year in the Americas.

Occurs from Senegal to Angola; also in the Americas.

Uses

The bark is used in tanning, for example to preserve fishing lines or nets and to colour leather dark red; bark infusion used in local medicines for toothache, for cleaning sores and application to pustules, and in decoctions against diarrhoea. The wood is hard and heavy, and is used in boat-building and construction; it also makes good charcoal. However, cut stumps do not regrow like other mangroves.

Conservation status

Widespread, least concern (LC) on a global basis.

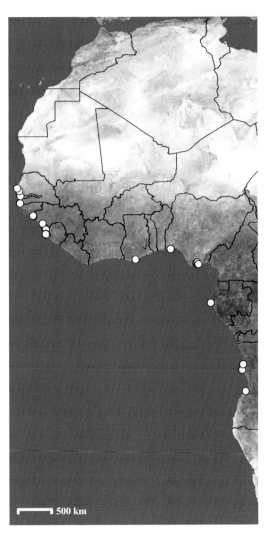

500 km

Common names:

Red mangrove; Manko (Mandinka, Gambia).

Rhizophora racemosa G.Mey.

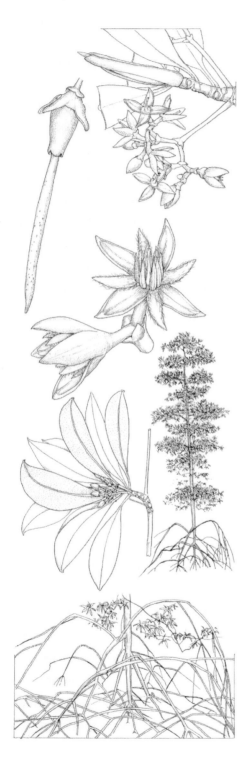

The commonest and largest Red Mangrove in West Africa. A pioneer on mud.

Field characteristics
All three *Rhizophora* are distinct by their stilt roots; this species has many flowers in short stout inflorescences.

Description
Tree 3.6–40 m tall. Trunk to 2.5 m in girth. Wood reddish, very hard, of close texture, brittle and gummy. **Bark** smooth, becoming furrowed with age. **Breathing roots** stilt roots abundant from the lower trunk, as well as slender aerial roots hanging from main branches. **Young twigs** swollen at nodes. **Leaves** narrowly elliptic, 7–15 × 2.5–6 cm, tip pointed; with reddish midrib, underneath with fine black cork dots when dry. **Flowers** many in much-branched clusters, inflorescence short and stout; flowers scented; buds bluntly pointed at tip. Sepals greenish. Corolla with 4 greenish-white hairy petals, deciduous. **Fruit** bottle-shaped, tough, with bracts (modified leaves) and swollen calyx at base, green, about 3.5–5 cm long, 1.8 cm across, dotted with lenticels; germinating on the tree with the root reaching 30–65 cm long while still on the tree.

Rhizophoraceae

Ecology and distribution

A pioneer on mud deposits, often forming large and pure stands, colonizing the seaward edge of the mangrove. This species often forms dense tangly thickets to 10 m high, full of stilt roots and aerial roots, and is very visible as it forms the outside edge of the mangrove vegetation.

Flowering times: in Nigeria AUG–DEC.

Fruiting times: most seasons.

Occurs from Senegal to Angola; also the Atlantic coast of tropical Americas.

Uses

The reddish, dense, hard and rather brittle wood is used in house building. The prop-roots are used to make racks for drying fish and walking-sticks. The bark and leaf-paste are rich in astringents and are used as toothpaste and in gargles for dental care.

Conservation status

Widespread, least concern (LC) on a global basis.

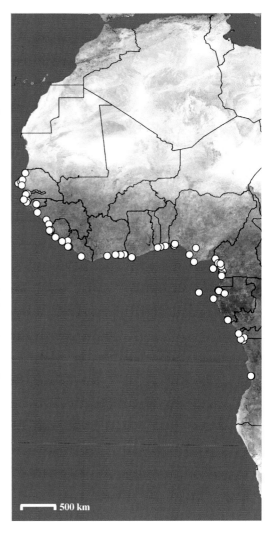

Common names:

Mangui (Wolof, Senegal), Tarrafe (Crioulo, Guinea Bissau), Séné (Balanta, Guinea Bissau), Akocut (T, Sierra Leone), Abuma, Kinsii (?Kinsu) (T, Sierra Leone), Oto (Togo), Egba (Nigeria, Lagos), Inetande, Ntame (Pahouin, Gabon), Inetanda (Gallois, Gabon).

Atlantic coast associate species

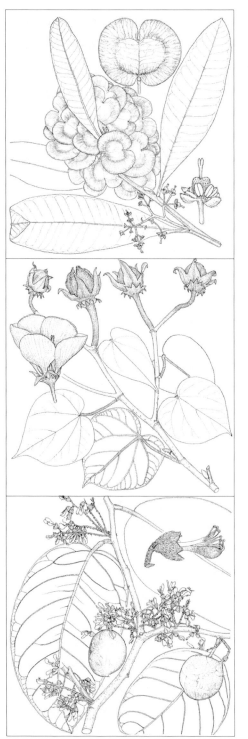

Dodonaea viscosa (Sapindaceae) — shrub to 3 m with alternate sticky leaves; at landward edge of mangrove and at slightly higher level. Widespread, also in East Africa. Wood used for fuel.

Hibiscus tiliaceus (Malvaceae) — small tree to 15 m with heart-shaped leaves; flowers large and yellow with a dark purple heart. Transition zone to fresh water; along creeks. Widespread; also in East Africa. Wood hard and durable, used in boat-building and cabinet-making.

Dalbergia ecastaphyllum (Leguminosae) — shrub to 6 m with purple twigs and alternate leaves 8–17 × 4–7 cm; white fragrant flowers. Landward side, occurs on drier soils. Senegal to Angola.

Trees and shrubs Alternate leaves

Scaevola plumieri (Goodeniaceae) — succulent low shrub with fleshy obovate leaves. Flowers in short clusters and a fleshy fruit. On sandy shores. Widespread, also in East Africa.

Terminalia catappa (Combretaceae) — tree to 24 m, with branches in 'layers'; leaves alternate, obovate, to 25 × 10 cm; fruit fleshy, ellipsoid, slightly winged. Landward side, occurs on drier soils. Native of India, naturalized along African coasts.

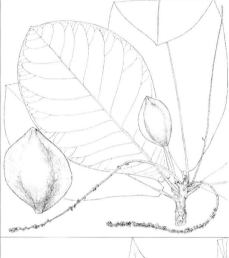

Thespesia populnea (Malvaceae) — tree or shrub with alternate leaves with cordate base, with minute scales on both surfaces; large yellow flowers; almost round fruits. Landward side bordering the mangrove vegetation. Widespread, also in East Africa.

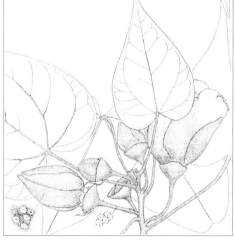

Cynometra megalophylla (Leguminosae) — tree to 20 m with leaves consisting of 2–4 pairs of leaflets; on sand banks on the landward side of the mangrove from Ghana to Nigeria.

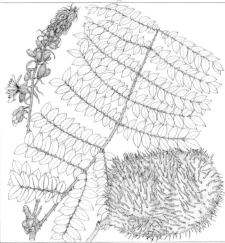

Guilandina bonduc (Leguminosae) — shrub or half-climbing tree to 5 m; stems with spiny prickles; leaves compound, with many small leaflets and recurved prickles. Flowers yellow. Fruit elliptic-flattened, with many close prickles; seeds ± globose, 2 cm across, hard, will float. Widespread on shores, may also occur in mangrove. A common drift seed. From Senegal to Angola; also in East Africa.

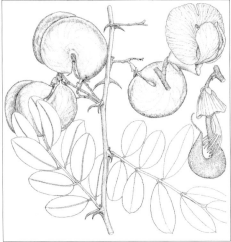

Drepanocarpus lunatus (Leguminosae) — thorny shrub, with recurved spines in pairs and leaves with 5–9 leaflets. Landward side, occurs on drier soils; on sand banks on the landward side of the mangrove. Senegal to Angola.

Oxystigma mannii (Leguminosae) can form almost monospecific stands fringing the mangrove in SW Cameroon (Cheek, 1992). Leaves with usually a single pair of leaflets. Flowers white or pink. Fruit irregular and knobbly.

Trees and shrubs Palms and pandans

Pandanus candelabrum (Pandanaceae) – 'screw pine', tree to 8 m with stilt roots and spiny trunk, and a crown of long spiny-margined leaves to 1.5 m long; in colonies, with *Rhizophora* in open lagoons. Senegal to Gabon, possibly further south.

Phoenix reclinata (Palmae) — palm to 10 m, usually with several stems from base; the lower leaflets grading into spines; transition zone to fresh water; widespread, also in East Africa.

Raphia vinifera (Palmae) — palm with leaves to 20 m long, usually with several stems from base; usually in freshwater swamps, but tolerates some salt; Ghana to Congo delta.

Herbs

Chenopodiaceae: **Arthrocnemon**, **Salicornia**, **Suaeda** — succulent herbs widespread along coasts, thrive in areas with high salt concentrations.

Sesuvium portulacastrum (Aizoaceae) — often rooting at the nodes, leaves succulent and obovate; common shore plant, also in dried out parts of lagoons, where flooding is irregular and salinity is very high through evaporation of water; widespread, also in East Africa. Young shoots are eaten.

Philoxerus vermicularis (Amaranthaceae) — small, rather succulent herb in sandy sites and dried-out channels from Senegal to Angola.

Pedalium murex (Pedaliaceae) — annual succulent-stemmed herb; leaves elliptic or obovate; fruit 4-angled with a spine on each angle. Coastal plant, salt indicator; widespread, also in East Africa.

Zygophyllum gaetulum var. **waterlotii** (Zygophyllaceae) — fleshy shrublet; leaves with 2 leaflets; fruits linear, hairy, 5-angled. Senegal.

Epiphytes

Epiphytic orchids may occur on mangrove trees; Cheek (1992) reports on a collection of nineteen orchid species made in a SW Cameroon mangrove forest during a single morning.

Mistletoes associated with mangroves include **Phragmanthera leonensis**, **Tapinanthus buchneri** and **T. sessilifolius**.

Grasses

Phragmites karka — perennial reed, in freshwater swamp fringing mangrove swamp. Widespread, also in East Africa.

Paspalum vaginatum — creeping perennial, common in areas with high salt content, e.g. between limits of high water mark and sping tides; pantropical.

Sporobolus spicatus — short tufted perennial, sometimes mat-forming; salt marsh, saline soils. Widespread, also in East Africa.

Sporobolus virginicus — spreading perennial, on highest ridges between the mangrove and the beach. Widespread, also in East Africa.

Sedges

Cyperus crassipes — common on sand. Widespread, also in East Africa.

Fimbristylis obtusifolia — tufted sedge with dense brown heads. Widespread, also in East Africa.

Kyllinga robusta — leafy sedge with dense reddish heads. Senegal to Congo.

Mariscus ligularis – coarse sedge with reddish heads. Senegal to Congo.

Ferns

Acrostichum aureum is a true mangrove fern, with leaves to 2 m; each leaf with up to 30 leaflets of which the upper may be fertile (spore-producing); may be very common at river mouths. Other ferns such as **Bolbitis auriculata** may occur on the landward edge of mangrove vegetation. Epiphytic ferns may be common on mangrove trees: **Polypodium**, **Nephrolepis**, **Platycerium**, **Asplenium africanum**, **Microsoria scolopendria**.

Box 2: Lower plants of the mangrove community

Decomposition takes place in the mangrove substrate. Aerobic and anaerobic bacteria and fungi enable the establishment of cyanobacteria; after this green algae and diatoms can be established. These are followed by lower animals such as nematodes and copepods, which further enrich the soil by secreting ammonia and phosphate. At this stage the soil is ready to support mangrove growth (Kalk 1995).

Chlorophyta (green algae) and Rhodophyta (red algae) are often present on silt and on the regularly inundated parts of the mangrove trees. Phaeophyta (brown algae) and some mat-forming Cyanophyta (blue-green algae) also occur in mangrove areas.

Seaweeds in mangrove habitats in East Africa include the green algae *Boodlea composita*, *Chaetomorpha crassa*, *Caulerpa scalpelliformis*, brown algae *Dictyota bartayresiana*, *D. cervicornis*, *Hidroclathrus clathratus* and red algae *Bostrychia tenella*, *Caloglossa leprieurii*, *Catanella caespitosa*, *Hypnea cornuta*, *Gracilaria edulis*, *Murrayella periclados*.

In the Niger Delta red algae include *Bostrychia binderi*, *B. calliptera*, *B. moritziana*, *B. radicans*, *Caloglossa leprieurii*, *C. ogasawaraensis*, *Catenella caespitosa*; green algae *Rhyzoclonium ambiguum*, *R. implexum*, *R. riparum* and *Cladophoropsis membranacea*.

Lawson (1985) pointed out a zonation pattern on aerial mangrove roots (*Avicennia* and *Rhizophora*) with the alga *Rhyzoclonium* forming a green zone above *Bostrychia* and *Caloglossa*; *Cladophoropsis* either occurs in the roots or forms a carpet over mud surface. A red alga *Bangia atropurpurea* was recorded on *Rhizophora* roots in Ikorodu, Nigeria. Species of *Brostrychia* and *Catenella* were also recorded in other countries in West Africa.

Mangrove mat-forming blue-green algae in Zanzibar include *Aphanocapsa* sp., *Lyngbya* sp., *Microcoleus* sp., *Nodularia* sp., *Oscillatoria* sp. and *Syctonema* sp. (Lugomela, 2002).

Epiphytic bryophytes for the Niger delta are treated by Odu (1985) and include *Calymperes erosum*, *C. palisotii*, *Octoblepharum albidum* (mosses) and *Ceratolejeunea* sp. (a liverwort). Species details for the Niger delta are given by Lawson (1985); for Kenya by Barnett & Briggs (1983); for Eastern Africa by Coppejans *et al.* (1997); for Tanzania by Jaasund (1976); for South Africa by Berjak *et al.* (1977); for Mozambique by Macnae & Kalk (1962), Kalk (1995), and Bandeira *et al.* (2001).

Avicennia marina flowers, northern Mozambique

Avicennia marina (Forssk.) Vierh.

Avicennia marina is the most widespread mangrove tree on the African and Madagascan Indian Ocean coasts, and also occurs along the Red Sea.

Field characteristics
Alternate leaves that are silvery hairy beneath, combined with many pencil-like breathing roots.

Description
Shrub or tree 1–15 m tall, with a spreading rounded crown. **Bark** flaky, brownish yellow-green to whitish-grey, in older trees with pink-brown spots; with aerial 'breathing' roots (not reaching the ground) on the lower portion. **Breathing roots** system extensive and shallow, without tap root but with large number of subterranean roots radiating outwards, 20–50 cm below soil surface; pneumatophores erect, simple, 5–60 cm high from subterranean roots, pointed, corky, lenticellate. **Young twigs** nodes swollen. **Leaves** opposite, elliptic or ovate, 3–12 × 1–5 cm, olive green above, matt and silvery-green hairy below, exuding salt. **Flowers** in small, dense heads, yellow to orange, fragrant. Corolla 4–7 mm long, 4-lobed. Stamens in the throat. **Fruit** subglobose to ovoid, green, 1–3 cm long, 0.7–2.5 cm in diameter, tomentose, 1-seeded, opening by 2(–4) valves on contact with water. Seed germinating within the fruit while still on the tree, then dispersed by tides but establishing rather quickly.

Avicenniaceae

Ecology and distribution

In sandier parts and inland fringes of mangrove association, mud of tidal rivers; colonizes new mud banks.

Flowering times: Eritrea JAN–MAR, SEP, DEC; Somalia JAN, MAY–OCT, DEC; Madagascar FEB–MAR, JUN, SEP–DEC.

Fruiting times: Madagascar JAN–MAR, OCT.

Occurs from Egypt to South Africa, also in Arabia, Asia and the west Pacific.

Uses

The wood is used in boat-building, carpentry and joinery, for poles and as fuel (high quality charcoal) over most of the distribution area. The bark supplies a brown dye, is used as animal fodder in dry areas and has minor medicinal uses, for example against fever, stomach and liver problems.

Conservation status

Widespread, least concern (LC) on a global basis.

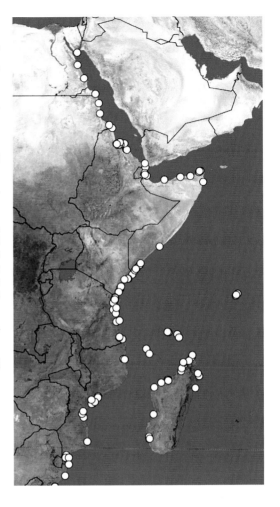

Common names:

Meshuz (Arab, Egypt), Showarab, Shawerab, Shora (Arab, Sudan), Sciavro (N Somalia); Mchu, Muchwii, Mtu (Swahili), Mtswi (Giriama, Kenya), Muchwi mume (Kiduruma), Mchu (Zigua, Tanzania), Thowozi, Utovoji, Unthoboti (Ronga/Thonga), Nhamesso (Bitonga), Musso, N'tsowozi, Txomahati, Invede, Mpedge, Mangal branco, Salgheiro (Mozambique), Fanavitraina, Tsingafiafy (Betsimisaraka, Madagascar), Afiafy, Hafihafy (Madagascar, W coast), Honkolahy, Mosotro, Masotry (Madagascar, NE), Honkofotsy (Sakalava, Madagascar).

Barringtonia asiatica (L.) Kurz

Sometimes known as *Barringtonia butonica* or *B. speciosa*

Not a true mangrove tree, as it lacks visible adaptations, but it often occurs on the edges of mangrove vegetation.

Field characteristics

A drift fruit on East African beaches; the large alternate leaves are distinctive.

Description

Evergreen shrub or tree to 30 m tall, without visible special adaptations; crown spreading. **Bark** rough and thick. **Breathing roots** none. **Young twigs** thick, 6–10 mm in diameter. **Leaves** alternate, tufted at ends of branches, shiny green, simple, obovate, 15–50 × 7–30 cm, hairless; marginal vein distinct. **Flowers** in small terminal groups of 3–20, white, 5–8 cm long, with 4 free petals and many stamens; opening in the evening, fragrant. **Fruit** top-shaped, 8–15 cm long, 4-angular, with fleshy-fibrous outer layer, single-seeded, indehiscent, will float.

Ecology and distribution

Littoral, on sandy or rocky beaches exposed to high tides.

Flowers open by night.

Flowering times: Madagascar Jan–Jun, Oct–Dec.

Fruiting times: Madagascar Apr, Oct–Dec.

Occurs in Zanzibar, Pemba; from Madagascar and the Indian Ocean Islands to Malaysia and Australia, Tahiti.

Uses

The bark is toxic to fish and is sometimes used in fishing. The seeds yield an oil formerly used in lamps.

Conservation status

Widespread, least concern (LC) on a global basis.

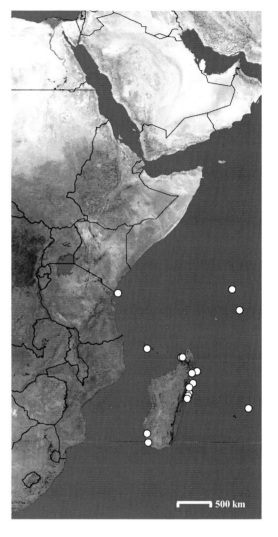

Common names:

Antsombera (Nosy Komba, Madagascar), Fotabe (Sambava, Madagascar), Fotaka (Toamasina, Madagascar).

Barringtonia racemosa (L.) Spreng.

Not a true mangrove tree, as it lacks visible adaptations apart from the occasional breathing root.

Field characteristics

Large alternate leaves can only be confused with *Barringtonia asiatica*; that species has entire leaf margins, while this one usually has toothed leaves.

Description

Tree 2–12(–27) m tall, with short and often twisted trunk, crown rounded, with spreading branches. **Bark** smooth or fissured, grey or yellowish. **Breathing roots** sometimes present as knobs or loops. **Young twigs** 4–9 mm diameter. **Leaves** in large clusters at branch ends, alternate, dark green, elliptic to obovate, 5–42 × 2–16 cm, hairless. **Flowers** many, in hanging clusters to 100 cm long, white, often tinged pink outside, 1–3 cm long, 4 free petals, many stamens; flowering at night, falling early next morning. **Fruit** reddish green, ellipsoid or conical, 3–9 cm long, 2–5 cm diameter, usually 1-seeded, very fibrous, dispersed by water. Seed ovoid or 4-sided, 2–4 cm long, 1–2.5 cm diameter, indehiscent, with fleshy-fibrous outer layer.

Lecythidaceae

Ecology and distribution

Landward edge of mangrove, also further inland. May form pure stands in swampy sites on the landward side of mangrove swamps.

Flowers open by night.

Flowering times: Madagascar MAY–JUN, AUG–JAN.

Fruiting times: Madagascar JAN–MAR, MAY–AUG, NOV.

Occurs from Somalia to South Africa, in Madagascar, and the Indian Ocean Islands to the Pacific.

Uses

The bark is used for cordage.

Conservation status

Widespread, least concern (LC) on a global basis.

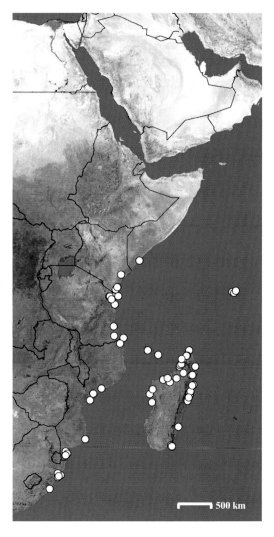

Common names:

Mtomondo (Swahili), Mtoro-toro, Mtovo-tovo (Namagoa, Mozambique), Madaroboka, Manondro, Mahondro, Manondra, Fota-be, Fotatra, Fotadrano, Mangaoka (Madagascar).

Bruguiera gymnorhiza (L.) Lam.*

A fairly common tree on the landward side of mangrove vegetation.

Field characteristics
Looks like a *Rhizophora*, but the buttresses and the absence of curved stilt roots should distinguish it.

Description
Shrub or tree 4–30 m tall, crown thick, branches opposite. Short buttresses may be present. **Bark** rough, dark reddish brown. **Breathing roots** radiating from trunk, numerous, often above soil; buttresses present; pneumatophores are knee roots, looping in and out of the soil and to about 10 cm high above it, lenticellate. **Young twigs** often covered with waxy bloom. **Leaves** appearing as a rosette but really opposite, glossy green above, often reddish beneath, hairless, rather leathery, elliptic, 5–22 × 2.5–10 cm; stipules in pairs. **Flowers** solitary and axillary, nodding, to 3.5 cm long. Calyx with 8–14 narrow lobes. Petals white, free, 8–14. Stamens many. **Fruit** top-shaped, crowned by calyx-lobes, to 4.5 cm long, 1-seeded, germinating on the tree with a spike-like seedling axis to 25 cm long emerging, and establishing quickly on contact with soil.

* The name often occurs with a spelling 'gymnorrhiza'. The original spelling by Linnaeus is with a single 'r', and though there have been many discussions over the correctness of the spelling, we believe that as long as these discussions have not reached a grammatical conclusion, we should stick with the original spelling (International Code for Botanical Nomenclature article 60.1).

Rhizophoraceae

Ecology and distribution

Less exposed parts of mangrove swamps; more tolerant of fresh water than other mangroves. May occur scattered in zones of *Rhizophora*, *Ceriops* and *Avicennia*, but prefers some shade for seedling establishment.

Flowering times: Madagascar Jan, Mar, May–Jun, Sep–Dec.

Fruiting times: Madagascar Jan–Feb, Jun, Sep, Dec.

Occurs from Eritrea to South Africa, and in Madagascar and Indian Ocean islands; also tropical Asia and Pacific.

Uses

The wood is hard and durable (when seasoned) and is used for building and construction, and high-quality charcoal and firewood. The bark is used for tanning throughout the distribution area.

Conservation status

Widespread, least concern (LC) on a global basis.

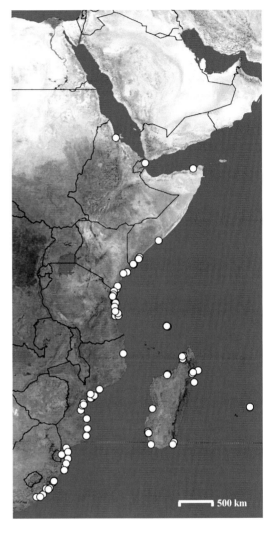

500 km

Common names:

Aysilola (Afar, Djibouti), Muia (Baguini, Somalia), Msinzi, Mchonga, Msisi, Mui, Muea (Swahili, Kenya & Tanzania), M'poronda, M'Finji (Mozambique), Vahona, Lavasiko, Tsitolonina, Tsitolona, Tanga foly ('fibrous bark'), (Madagascar).

Bruguiera gymnorhiza, Inhaca island, Mozambique

Bruguiera gymnorhiza flower

Mangrove in Mabeta-Moliwe, Cameroon

Ceriops tagal (Perr.) C.B.Robinson

Sometimes called *Ceriops boiviniana, C. somalensis*

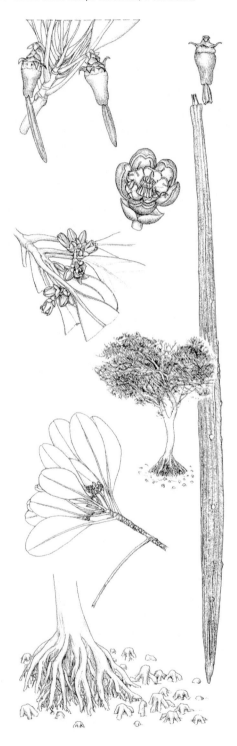

A common mangrove tree, *Ceriops* prefers the landward side of mangrove vegetation, where it may form pure stands.

Field characteristics

The knee-shaped breathing roots are distinctive; *Bruguiera* has these as well but that species has rough bark.

Description

Shrub or tree to 6 m tall, evergreen, crown compact, small buttresses may be present. **Bark** smooth, light grey or brownish yellow. **Breathing roots** system with radiating roots from trunk, forming small buttresses on the trunk, and looping in and out of the soil to form small knee roots; stilt roots sometimes present. **Young twigs** jointed, with swollen nodes. **Leaves** in clusters, opposite, pale yellowish green, elliptic to obovate, 3–10 × 1–5 cm. **Flowers** in short axillary clusters, white, small. Calyx 5–6-lobed. Petals 5. Stamens 10–12. **Fruit** ovoid, 1.5–2.5 cm long, usually 1-seeded, germinating on the tree, seedling to 30 cm long and angular.

Rhizophoraceae

Ecology and distribution

Locally common on the shoreward side of mangrove swamps; prefers drier areas; may form narrow bands along channels where sand has accumulated on the banks. Seedlings may require some shade for establishment.

Flowering times: Madagascar JAN–JUN, AUG, OCT–DEC.

Fruiting times: Madagascar JAN, MAR–JUN, AUG–NOV.

Occurs from Djibouti to South Africa, Madagascar, and Asia to New Guinea.

Uses

Good for firewood and charcoal; used for poles, for fishing trap stakes and for boat-building; the bark is used in dyes and tanning.

Conservation status

Widespread, least concern (LC) on a global basis.

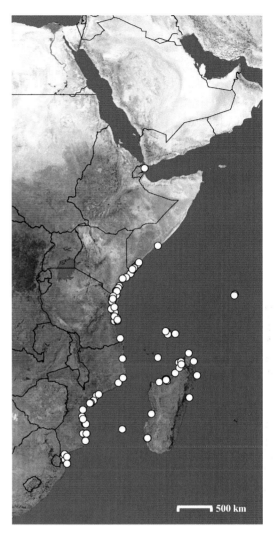

Common names:

Ginni Kandala (Djibouti), Sciori Medu (Somali), Mianda (Baguini, Somalia), Mkandaa, Mkandaa mwekendu, Mkandaa ya pwani (Swahili, Kenya & Tanzania), Lichochochana (Inhaca, Mozambique), Hlothlotxwani (Loaga, Mozambique), Kapa, Cava (Macua, Mozambique), Honkolahy, Honkovavy, Farafaka (Madagascar).

Heritiera littoralis Ait.

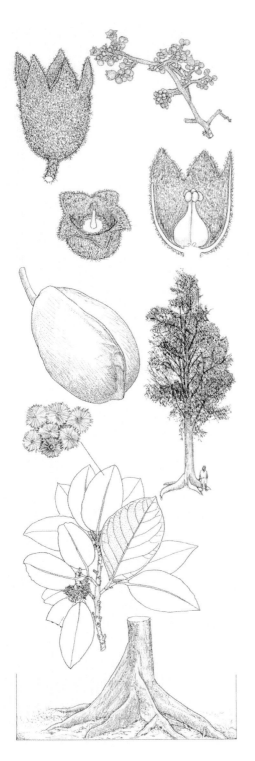

Heritiera is a common tree on the landward side of the mangrove.

Field characteristics
Easily distinguished by the silvery-scaly undersides of the leaves and the ribbon-like surface roots. The ridged fruit is also characteristic.

Description
Evergreen tree to 21 m tall, with thin buttresses. **Bark** grey, fissured. **Breathing roots** ribbon-like surface roots. **Young twigs** with silvery scales. **Leaves** alternate, with silvery scales underneath, 10–20 × 5–12 cm, stipules in pairs, deciduous. **Flowers** yellow-green, in axillary stalked clusters, 4–5 mm long and without petals, either male or female. **Fruit** shiny brown with 1–4 partial fruits, each 6–8 × 3–6 cm, ellipsoid, flattened and with a ridge on one side, dispersed as drift fruits.

Sterculiaceae

Ecology and distribution Landward side of mangrove, also scattered on raised sandy banks along channels, among *Ceriops* and *Lumnitzera*; not an obligatory mangrove, as it occurs in other coastal vegetation types as well. The fruit always floats with the ridge upwards, and this ridge acts as a sail!

Flowering/fruiting times: not enough data.

Occurs in Kenya, Tanzania, Mozambique, Madagascar, tropical Asia, Australia and the Pacific.

Uses
A source of charcoal, also used for poles and masts of boats.

Conservation status
Widespread, least concern (LC) on a global basis.

Common names:

Msikundazi (Swahili, Tanzania), Moromony (Madagascar, Ambanja).

Lumnitzera racemosa Willd.

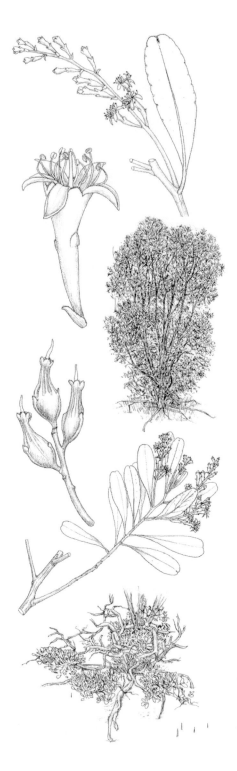

Lumnitzera grows on the landward side of mangrove vegetation.

Field characteristics
The radiating surface roots and alternate leaves.

Description
Shrub or tree to 9 m tall, deciduous. **Bark** rough, fissured, reddish-brown. **Breathing roots** system of radiating roots spreading outwards from trunk, with angular buttresses; side roots sometimes looping in and out of the soil to form knee roots. **Young twigs** reddish-brown, with close leaf scars, hairless. **Leaves** spirally arranged, fleshy, narrowly obovate to elliptic, 2–8 × 1–3 cm, tip rounded, hairless. **Flowers** in axillary spikes to 2 cm, white or cream, 7–8 mm long, with 5 free petals and 10 stamens. **Fruit** woody, tapered to both ends, 10–12 mm long, 3–5 mm in diameter, 1-seeded, spreading as a drift fruit.

Combretaceae

Ecology and distribution

Landward side of mangrove at about high water mark; along small channels with sandy banks, together with *Ceriops*.

Flowering times: seems to flower most months in Tanzania and Madagascar.

Fruiting times: not enough data.

Occurs from Kenya to South Africa, in Madagascar, and tropical Asia to Polynesia.

Uses

Good for fuel (in Kenya and Tanzania) and firewood, used as poles for building (Mozambique).

Conservation status

Widespread, least concern (LC) on a global basis.

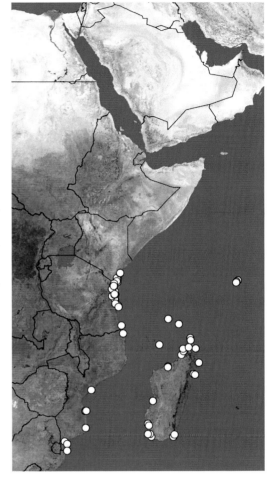

Common names:

Mkandaa, Kikandaa, Mkandaa Mwitu, Mkandaa-dume (Swahili, Kenya), Mkaracha ya pwani (Mafia, Tanzania), Vahona, Votishonko, Lovingo, Rongo, Sosuman (Madagascar).

Pemphis acidula Forst.

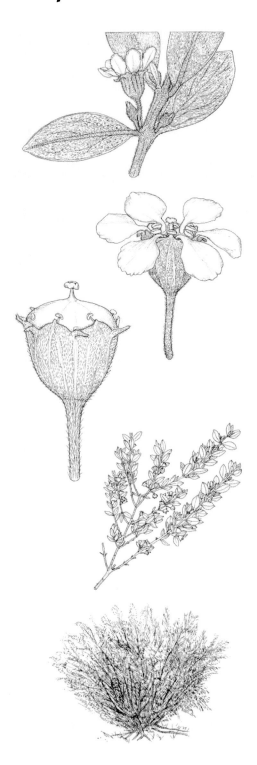

Not a true mangrove tree, as it lacks visible adaptations, but a common mangrove associate.

Field characteristics
Twigs and opposite leaves with small silvery hairs; fruit with persistent calyx.

Description
Shrub or tree to 11 m tall, but occasionally in very poor conditions a dwarf shrub. **Bark** grey, rough and flaking, in older trees deeply fissured. **Breathing roots** none. **Young twigs** angular, with silky silvery hairs. **Leaves** opposite to almost whorled, leathery to fleshy, oblanceolate to lanceolate, 0.8–3.5 × 0.3–1.3 cm, tip blunt or nearly so, with small silvery hairs on both sides. **Flowers** in upper axils, solitary or rarely in pairs, white or pink. 6 petals , 5 mm long, falling soon. Stamens 12. **Fruit** purple, subglobose, ± included in the calyx, 4–5 mm in diameter, with persistent style.

Lythraceae

Ecology and distribution

Landward side of mangrove at or below high water mark, also on dunes and in coral rag thicket.

Flowering times: most months in Tanzania, with a peak in Ocт.

Fruiting times: not enough data.

Occurs from Kenya to Mozambique, in Madagascar, and the Indian Ocean Islands to Pacific Ocean.

Uses

A source of firewood, charcoal and building wood.

Conservation status

Widespread, least concern (LC) on a global basis.

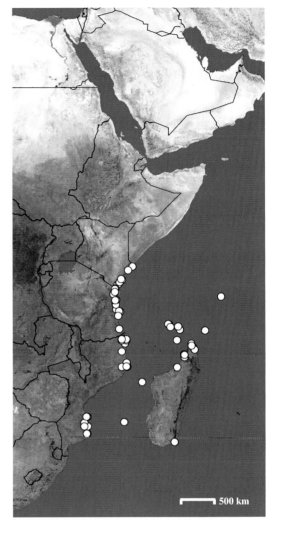

500 km

Common names:

Kilalamba kike, Mkaa pwani, Mkamkam, Mnyinyuwa, Mwanawea mpwa (Swahili), Musisi (Giriama), Sukire (Macua), Gama, M'gama, Megama, Muvucuta (Quimuase), Vonjioko (Malagasy), bois d'amande (Indian Ocean Islands).

Rhizophora mucronata Lam.

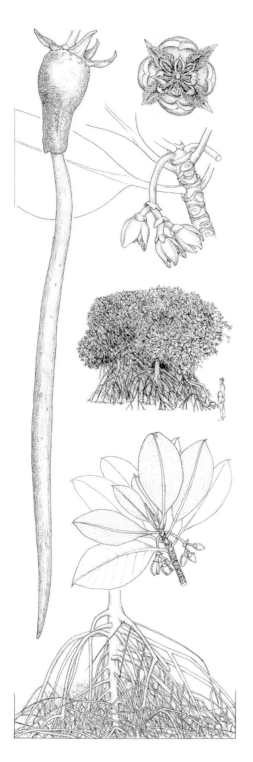

This *Rhizophora* is the only one on the east coast of Africa, and is a very common species of creeks and channels.

Field characteristics
Distinguished by arched stilt roots; and its fruit which has a long seed-sprout while still on the tree.

Description
Tree to 25 m tall, hairless, with opposite branches. Crown dense. **Bark** fissured, dark. **Breathing roots** system extensive, with stilt roots emerging from the lower trunk and curving outwards, lenticellate just above soil surface, going into the soil, and there sending out small secondary roots; sometimes aerial roots present, growing down from branches. **Young twigs** 3–9 mm in diameter. **Leaves** in clusters, opposite, elliptic, 8–18 × 4–9 cm, the tip with characteristic sharp bristle; underneath with fine black cork dots when dry. **Flowers** in short axillary clusters of 4 or more, cream, to 1 cm long. Calyx 4-lobed. Petals 4. Stamens 8. **Fruit** pear-shaped, 3–4 cm long, leathery, 1–seeded, germinating on the tree with a spike-like rootlet to 45 cm emerging, the whole embryo falling out of the fruit and dispersing by water.

Ecology and distribution

Common on the seaward side on mud deposits; often the dominant species in brackish water at river mouths.

Flowering times: Tanzania Jan–Mar, Jun–Aug, Oct–Dec.

Fruiting times: Tanzania Jan–Mar, Jul–Aug, Nov–Dec.

Occurs from Eritrea to South Africa, Madagascar, and the tropical Old World.

Uses

The sap and bark are used for tanning (for example of nets and fishing lines), dyes and as fish poison. The wood is used for fuelwood, high quality charcoal and building poles; the split stems are used for grain-sifting baskets; the stilt roots are used for fishing trap stakes.

Conservation status

Widespread, least concern (LC) on a global basis.

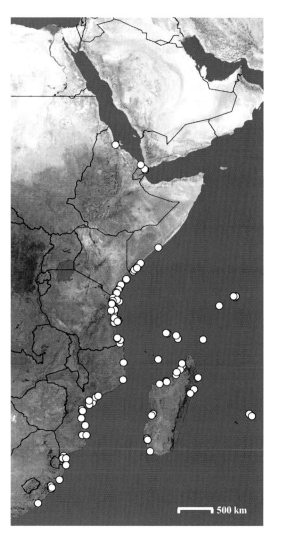

Common names:

Mucoco, Msindi (Bagiuni, Somalia); Kandalo, Gandel, Sciori Guduo (Somali); Mkoko, Mgando, Mliana, Msinzi (Swahili, Duruma, Giriama, Kenya, Tanzania); Mtongakima (Tanzania, Mafia); M'Fingi, M'Candalla (Mozambique), Xinkaha, Shinkanga (Ihonga/Ronga, Mozambique), Mtulo (Macua, Mozambique); Honkovavy, Honkolahy (Madagascar).

Rhizophora mucronata, northern Mozambique

Rhizophora mucronata, northern Mozambique

Rhizophora mucronata seedlings, Luchete, northern Mozambique

Sonneratia alba Sm.

Sonneratia is a common tree of the seaward side of mangrove formations.

Field characteristics
The finger-like breathing roots and absence of stilt roots.

Description
Evergreen shrub or tree 3–20 m tall. **Bark** dark grey or grey-brown, rough and fissured. **Breathing roots** finger-like to conical, erect, many, to 75 cm high, rarely to 25 cm across. **Young twigs** cylindrical, thickened at the nodes, 3–6 mm in diameter. **Leaves** opposite, fleshy, without stipules, ovate or obovate, yellow-green, 4–12 × 2–9 cm, rounded or notched at tip, hairless. **Flowers** solitary or up to 3 together at shoot tip, scented, with 6–8 calyx lobes that are magenta-pink inside, 1.2–2 cm long, and as many inconspicuous white or magenta petals. Stamens many, exserted. **Fruit** with persistent calyx, obconic or top-shaped, 2–3 cm long, 3–4 cm in diameter.

Ecology and distribution

Seaward fringe of mangrove formations, on sand or mud; may be common or dominant, especially in deepest water.

Flowering times: Madagascar FEB–JUN, SEP–NOV.

Fruiting times: Madagascar JAN, SEP–NOV.

Occurs in East Africa from Somalia to Mozambique, Madagascar and the Indian Ocean Islands to the Pacific.

Uses

The light wood is used for carpentry and fuel, poles, dugout canoes, boat ribs, masts and paddles; the pencil roots are used as floats for fishing nets.

Conservation status

Widespread, least concern (LC) on a global basis.

Common names:

Mlilana (Bagiune, Somalia); Mlilana, also Mpira ('top', for the fruit), Mkoko mpia, Mkapa (Swahili, Kenya & Tanzania); Metinindi (Mozambique); Farafaka, Farahafaka, Vahona, Songery (Madagascar).

Large *Sonneratia alba*, northern Mozambique

Sonneratia alba, pneumataphores

Sonneratia alba, Mocímboa da Praia, Mozambique

Xylocarpus granatum Koen

Sometimes called *Xylocarpus benadirensis*

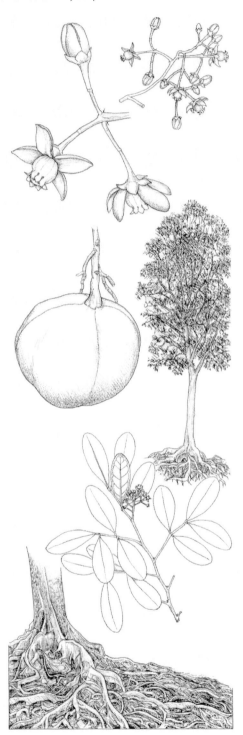

Xylocarpus is a widespread species of the landward side of mangrove formations.

Field characteristics

The compound leaves are unique among true mangrove species; the fruit is also unmistakable.

Description

Tree, 3–15 m tall, evergreen, with buttresses. **Bark** smooth, pale green or yellow, peeling in patches. **Breathing roots** ribbon-like surface roots emerging from the buttresses. **Young twigs** cylindrical, hairless. **Leaves** alternate, compound and consisting of 2–6 elliptic or obovate leaflets to 12 × 6 cm. **Flowers** small, in short axillary clusters to 6 cm long, 3–5 mm across. Calyx lobes 4. Petals 4, white. **Fruit** subglobose, leathery, 14–25 cm across, late-opening, with 8–20 corky-covered seeds which germinate in the fruit or soon after release.

Ecology and distribution

On tidal mud, on landward side of mangrove.

Flowering times: not enough data.

Fruiting times: not enough data.

Occurs from Somalia to Mozambique, Madagascar, Indian Ocean islands and Asia to Polynesia.

Uses

The wood is very hard, and has been used for dhow building, dugout canoes, boat-building, cart construction, furniture and buildings. Makes good firewood and charcoal, for example for smoking fish. Has minor medicinal uses in aphrodisiacs, against stomach-ache.

Conservation status

Widespread, least concern (LC) on a global basis.

NOTES. The related *X. moluccensis* (Lam.) Roem. is treated under associate species. It differs in the more ovate leaves with subacuminate tip (not obtuse to notched as in *X. granatum*) and the much smaller fruit; this species also lacks buttresses and surface ribbon-like roots.

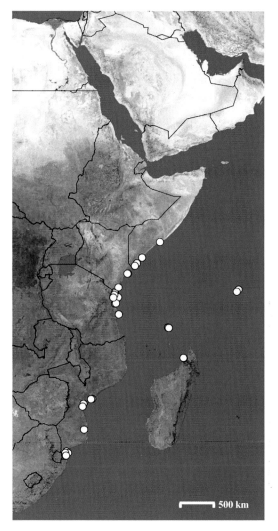

Common names:

Mucomafi (Bagiuni, Somalia); Mkomafi, Mtonga, Mkomasi (Swahili, Kenya & Tanzania); Seane, M'fava (Shizonga, Mozambique).

Xylocarpus granatum, bark, fruit and seed (counterclockwise from upper right)

Mangrove vegetation with *Xylocarpus*, near Mombasa, Kenya

Indian Ocean coast associate species

Brexia madagascariensis (Brexiaceae/ Grossulariaceae) — shrub or small tree to 10 m; leaves alternate, oblong to obovate, 3–35 × 2–8 cm. Flowers thick, whitish, 1–2 cm long, in axillary stalked umbels. Fruit 5-ribbed, 4–10 cm long. Tanzania, Mozambique, Madagascar; in mangrove edges.

Dodonaea viscosa (Sapindaceae) — shrub to 3 m with alternate sticky leaves; at landward edge of mangrove and slightly higher level. Widespread; also in West Africa. Wood used for fuel.

Foetidia obliqua (Lecythidaceae) — shrub or small tree; leaves alternate, elliptic to slightly obovate, 3–11 × 1–5 cm, with many close side-veins. Flowers in dense groups at branch tips, lacking a corolla, with many stamens. Fruit red, obconic, 1–1.5 cm long, with 3 long lobes. Pemba Island, Madagascar; presumably in or near mangrove.

Trees and shrubs Alternate leaves

Hibiscus tiliaceus (Malvaceae) — small tree to 15 m with heart-shaped leaves; flowers large and yellow with a dark purple heart. Transition zone to fresh water; along creeks. Widespread; also in West Africa. Wood hard and durable, used in boat-building and cabinet-making.

Terminalia catappa (Combretaceae) — tree to 24 m, with branches in 'tiers'; leaves alternate, obovate, to 25 × 10 cm; fruit fleshy, ellipsoid, slightly winged, will float. Landward side, occurs on drier soils; native of India, naturalized along African coasts.

Thespesia populnea (Malvaceae) — tree or shrub with alternate leaves with cordate base, with minute scales on both surfaces; large yellow flowers; almost globose fruits. Landward side bordering the mangrove vegetation; widespread, also in West Africa.

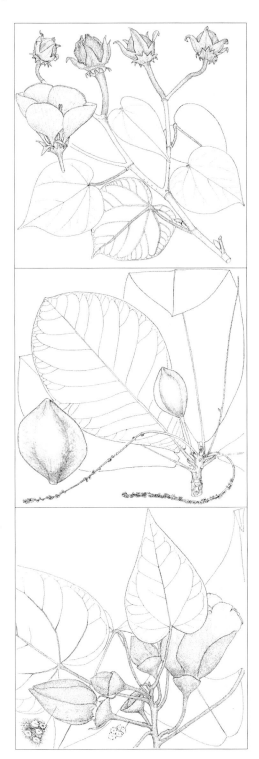

Trees and shrubs Alternate leaves

Scaevola plumieri (Goodeniaceae) — succulent low shrub with fleshy obovate leaves. Flowers in short clusters and a fleshy fruit. On sandy shores. Widespread, also in West Africa.

Trees and shrubs Opposite leaves

Calophyllum inophyllum (Guttiferae) — tree to 30 m; latex present in all parts; leaves opposite, elliptic, 10–20 × 5–12 cm, with many closely parallel veins. Flowers white, with many yellow stamens. Fruit globose, 3–4 cm across. Kenya to Mozambique and Madagascar, on sandy shores.

Trees and shrubs Compound leaves

Derris trifoliata (Leguminosae) — woody climber to 15 m long, young stems dark red; leaves with 3–7 leaflets; flowers white to pink; fruit a flat legume to 4 cm. On beaches, also in mangrove; widespread along Indian Ocean shores.

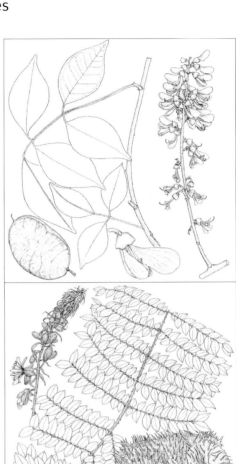

Guilandina bonduc (Leguminosae) — shrub or half-climbing tree to 5 m; stems with spiny prickles; leaves compound, with many small leaflets and recurved prickles. Flowers yellow. Fruit elliptic-flattened, with many close prickles; seeds ± globose, 2 cm across, hard, will float. Widespread on shores, may also occur in mangrove. A common drift seed. Also in West Africa.

Xylocarpus moluccensis (Meliaceae) — tree to 15 m with compound leaves with 2–6 leaflets; flowers in lax sprays; fruit round, to 8 cm across. Sandy or rocky beaches, Kenya to Mozambique, Madagascar and Indian Ocean islands.

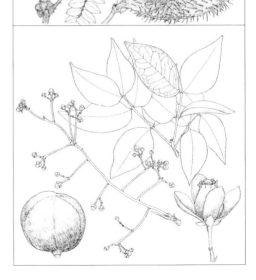

Hibiscus and *Pandanus* in Mabeta-Moliwe, Cameroon

Palms and pandans

Pandanus livingstonianus (Pandanaceae) — tree to 20 m with stilt roots, spiny trunk and tufts of long leaves on short side branches; transition zone to fresh water; Mozambique.

Phoenix reclinata (Palmae) — palm to 10 m, usually with several stems from base; the lower leaflets grading into spines; transition zone to fresh water; widespread, also in West Africa.

Herbs

Chenopodiaceae: **Arthrocnemon**, **Salicornia**, **Suaeda**; succulent herbs widespread along sea-coasts, thrive in areas with high salt concentrations.

Sesuvium portulacastrum (Aizoaceae) — succulent herb, often rooting at the nodes, leaves succulent and obovate; common shore plant, also in dried out parts of lagoons, where flooding is irregular and salinity is very high through evaporation of water; widespread, also in West Africa. Young shoots are eaten.

Trianthema triquetra (Aizoaceae) — creeping annual, much branched; leaves linear, succulent, 2 cm long. Flowers and fruit small. Salty sands and silts; widespread, also in West Africa.

Pedalium murex (Pedaliaceae) — annual succulent-stemmed herb; leaves elliptic or obovate; fruit 4-angled with a spine on each angle. Coastal plant, salt indicator; widespread, also in West Africa.

Cressa cretica (Convolvulaceae) — subshrubby herb, leaves ovate, <1 cm; flowers 5 mm long; fruit 3–4 mm. Mangrove and beach; widespread, also in West Africa.

Epiphytes

Epiphytic mistletoes (Viscaceae, Loranthaceae) occur on mangrove trees, e.g. **Viscum nervosum**, **Tapinanthus zanzibarensis**, **Erianthemum dregei**, **Oncocalyx bolusii**.

Epiphytic orchids may occur on mangrove trees but details are lacking as to which species.

Grasses

Dactyloctenium geminatum — mat-forming perennial with tough wiry stems; common near shore. From Somalia to South Africa.

Paspalum vaginatum — creeping perennial, common in areas with high salt content, e.g. between limits of high water mark and spring tides; pantropical.

Phragmites karka — perennial reed; freshwater swamp fringing mangrove swamp. Widespread, also in West Africa.

Sporobolus kentrophyllus — tussocky perennial; sandy patches within mangrove. Somalia to Tanzania.

Sporobolus spicatus — short tufted perennial, sometimes mat-forming; salt marsh, saline soils. Widespread, also in West Africa.

Sporobolus virginicus — spreading perennial; highest ridges between the mangrove and the beach. Widespread, also in West Africa.

Sedges

Cyperus crassipes — common on sand. Widespread, also in West Africa.

Fimbristylis obtusifolia — tufted sedge. Widespread, also in West Africa.

Rushes

Juncus kraussii — rush to 1.5 m high in low-salinity saltmarsh; may form large colonies.

Ferns

Acrostichum aureum is a true mangrove fern, with leaves to 2 m; each leaf with up to 30 leaflets of which the upper may be fertile (spore-producing); may be very common at river mouths.

Epiphytic ferns may be common on mangroves trees: **Polypodium**, **Nephrolepis**, **Platycerium**.

Sea-grasses may grow underwater on sandbanks and mudflats on the outside of the mangrove. These include **Nanozostera capensis** (Zosteraceae); **Cymodocea rotundata** (Cymodoceaceae), other Cymodoceaceae and Hydrocharitaceae such as **Thalassia hemprichii**.

A number of small amphibious fish, crabs and molluscs occur in the mud of the mangrove. They can tolerate the absence of water for brief periods. Species include mud-skippers (*Periophthalmus*), fiddler crabs (*Uca inversa, U. annulipes, U. gaimardi, U. urvillei, U. vocans*) marsh crab (*Sesarma catenata, S. eulimene, S. ortmanni*), mud crabs (*Scylla serrata, Ilyograpsus* species), whelks (*Cerithidea decollata, Terebralia palustris*), mussels, periwinkles (*Littorina scabra, Tympanotomus, Pachymelania*), oysters (*Ostrea tulipa, Crassostrea cucullata, Saccostrea cucullata*) and barnacles (*Balanus amphitrite*) (Berjak *et al.*, 1977; Barnett & Briggs, 1983; Moses, 1985; Powell, 1985).

At Inhaca Island (southern Mozambique), more than fifty animal species have been recorded from the mangrove vegetation. Half of these are crustaceans, the rest being mostly snails and fish (Kalk, 1995). In East Africa, 117 molluscs and 163 species of crustraceans have been recorded in mangrove vegetation (Taylor *et al.*, 2003).

Shrimps (for example *Penaeus duorarum*) and fish (anchovy, tilapia, snappers, shinynose, drum, and juveniles of several more taxa) occur as casual visitors, but some like the mullet (*Mugulidae*) are more regular inhabitants in open water within the mangrove vegetation, either in channels, or during the high tide. Taylor *et al.* (2003) report that 114 species of fish have been recorded in East African mangrove.

Insects include obligatory mangrove species like *Halobates*, a 'pond-skimmer', but most mangrove insects are facultative, which means that they are not strictly bound to the mangrove vegetation. Mosquitoes breed happily and in large numbers in shallow and stagnant water, including brackish water of the mangrove zone. *Aedes pembaensis* lays its eggs on the claws of the crab *Neosarmatium meinerti* (Hogarth 1999). Bees (*Apis* and *Xylocopa* species) commonly feed in the mangrove. Flies are particularly numerous, especially biting flies (sand flies, Psychodidae; horse flies, Tabanidae). Dragonflies, damselflies, beetles, dipterans and ants (including *Oecophylla*) as well as wolf-spiders (Lycosidae) and jumping spiders (Salticidae) are reported in Barnett & Briggs (1983).

The canopy is inhabited by many species, but few of these are restricted to the mangrove. The crab *Sesarma leptosoma* lives on mangroves, never entering the water or even venturing onto the mud, but feeding on mangrove root and leaf tissue (Vannini & Ruwa, 1994).

Common birds in mangroves, most of them seen in the canopy, include *Actites hypoleucos* (common sandpiper), *Ardea cinerea* (grey heron), *Ardea melanocephala* (black-headed heron), *Batis molitor* (white-flanked flycatcher), *Bubulcus ibis* (cattle egret), *Burhinus vermiculatus* (water thick-knee), *Butorides striatus* (green-backed heron), *Ceryle rudis* (pied kingfisher), *Cynnirys bifasciatus* (purple-banded sunbird), *Corythornis cristata* (malachite kingfisher), *Egretta alba* (great white egret), *E. garzetta* (little egret), *Gypohierax angolensis* (palm-nut vulture), *Halcyon senegaloides* (mangrove kingfisher), *Haliaeetus vocifer* (African fish eagle), *Lunchura cocullata* (bronze mannikin), *Nycticorax nicticorax* (black-crowned night heron), *Platysteira peltata* (wattle-eyed flycatcher), *Phalacrocorax africanus* (long-tailed cormorant), *Phoenicopterus ruber* (greater flamingo), *Ploceus xanthops* (golden weaver), *Pycnonotus barbatus* (black-eyed bulbul), *Serinus mozambicus* (yellow-eyed canary), *Streptopelia* spp. (turtle doves) and *Zosterops senegalensis* (yellow white-eye) (Berjak *et al.* 1977, Bennun *et al.* 1997). The endangered Madagascar teal, *Anas bernieri*, nests in holes in mangrove trees (Taylor *et al.*, 2003).

The mangrove forest of the Rufiji Delta (Tanzania) is an important site for migratory wetland birds, such as the curlew sandpiper (*Calidris ferruginea*), little stint (*Calidris minuta*), crab plover (*Dromas ardeola*), roseate tern (*Sterna dougallii*) and Caspian tern (*Hydroprogne caspia*) (Semesi, 1993).

In Madagascar, the mangroves in the northwest (Sakalava region) are the preferred habitat for the lemur *Propithecus verreauxi coronatus* (Roger & Andrianasolo, 2003). In East Africa, the Nile crocodile (*Crocodilus niloticus*), hippopotamus (*Hippopotamus amphibius*) and otter (*Lutra* species like *L. maculicollis*) are common, as well as various species of monkey (for example Sykes monkey, *Cercopithecus mitis*).

The mangroves of the Bazaruto Archipelago (Mozambique) support a large number of threatened and endangered turtles and marine mammals, and are particularly important for the last viable population of dugongs (*Dugong dugon*) in Eastern Africa. Sea turtles that visit are the endangered green turtle (*Chelonia mydas*), the critically endangered hawksbill turtle (*Eretmochelys imbricata*), and the endangered olive ridley (*Lepidochelys olivacea*). These species also nest around the river mouths of some of the larger mangrove stands in the region, particularly in northern Kenya, southern Tanzania and Mozambique (Hughes and Hughes, 1992).

The Zambezi delta in Mozambique supports a huge prawn fishery just offshore and in the vicinity of the mangrove stands there are many large sharks, a humpback whale nursery, and notable populations of porpoises. The mangroves themselves provide an important stopover site for migratory wetland birds.

Heritiera litoralis, near Gazi, Kenya

Glossary

Aerial roots — roots above the soil/mud; either from underneath the surface, or coming from trunk/branches

Anaerobic — without oxygen

Axillary — in the narrow angle between the stem and the stalk of a leaf

Blade — the expanded part of a leaf (not the stalk)

Buttress — a woody fin at base of tree increasing stability

Calyx — the outermost whorl of flower organs, usually divided into sepals

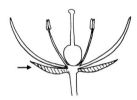

Compound — not simple; of a leaf, said when leaf consists of several leaflets

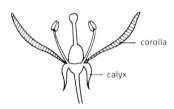

Corolla — the second whorl of flower organs, inside the calyx and outside the stamens

Crown — the upper, leafy part of a tree

Deciduous — leaves (or other parts) falling off, not evergreen

Drift seed/fruit — where the fruit/seed floats on the sea and is dispersed in that way

Ellipsoid — a 3-dimensional shape, elliptic in the vertical plane (broadest at the middle with 2 rounded ends)

Finger root — an erect breathing root with a narrowly cylindrical or conical shape

Fissured — (of bark) with longitudinal grooves or cracks

Globose — spherical, round (in three dimensions, like a ball)

Habit — the general appearance of a plant species

Habitat — the type of environment in which a plant grows

Indehiscent — (of fruit) not splitting open

Inflorescence — the part of the plant that bears the flowers

Jointed — with a joint, like a break

Knee roots — pneumatophores with a shape similar to a loop

Lanceolate — shaped like a spear-blade, pointed at both ends and with the widest part below the middle

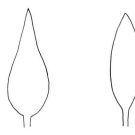

Lenticel — corky protuberance on the bark, allowing gas exchange

Lenticellate — with scattered lenticels

Obconic — a 3-dimensional shape, shaped like a child's top, the widest part near the top

Oblanceolate — shaped like a spear-blade, pointed at both ends and with the widest part above the middle

Obovate — shaped like an upside-down egg (the wider part near the top) (2-dimensional)

Ovoid — shaped like an egg (3-dimensional)

Pear-shaped — (see image)

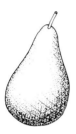

Peg roots — erect breathing root with a narrow conical shape, narrowest at the top

Pencil roots — erect breathing root with a narrowly cylindrical shape

Petal — flower lobe (part of the corolla)

Petiole — leaf-stalk

Pneumatophore — breathing roots raised above the soil/mud

Ribbon-like roots — shallow horizontal roots at the surface of the soil or mud

Sepals — the separate parts, or lobes, of the outermost whorl of floral organs, the calyx

Species — the basic unit of plant classification; plants looking very similar to each other and capable of interbreeding

Spike — an inflorescence where the flowers are sessile on an unbranched axis

Stamens — male organ of the flower

Stilt roots — lateral roots coming from the lower part of the stem and reaching the ground; also called prop roots

Stipule — a small appendage at the base of a leaf, usually on the stem

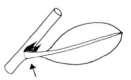

Style — stalk on top of the female organ of the flower (and sometimes later on the fruit)

Terminal — at the upper end

Viviparous — bearing live offspring, that is with seeds that sprout while still on the parent plant

Zonation — sequence of vegetation types in three dimensions (but not in time)

References

Bandeira, S.O., António, C.M. & Critchley, A.T. (2001). A taxonomic listing, including new distribution records, for benthic, intertidal seaweeds from Mecúfi, northern Mozambique. *S. African J. Bot.* 67: 492–496

Barbosa, F.M.A., Cuambe, C.C. & Bandeira, S.O. (2001). Status and distribution of mangroves in Mozambique. *S. African J. Bot.* 67: 393–398

Barnett, R.J. & Briggs, J.D. (eds) (1983). *University of Bristol Kenya Expedition 1982 Report.* Privately published by the Expedition members

Bennun, L., Knott, H. & Richmond, M.D. (1997). Class Aves. Coastal Birds. In: *A Guide to the Seashores of Eastern Africa and the Western Indian Ocean Islands*, ed. M.D. Richmond, pp. 372–383. SIDA, Department for Research Cooperation, SAREC, Stockholm

Berjak, P., Campbell, G.K., Huckett, I. & Pammenter, N.W. (1977). *In the Mangroves of Southern Africa.* Wildlife Society of Southern Africa, Natal

Cheek, M. (1992). *Botanical Survey of the Proposed Mabeta-Moliwe Forest Reserve in S.W. Cameroon.* Forestry Department, Government of Cameroon

Cheek, M. & Dorr, L. (2007). Flora of Tropical East Africa: Sterculiaceae. Royal Botanic Gardens, Kew.

Colloty, B.M. (2000). 'Botanical importance of estuaries of the former Ciskei/Transkei region'. PhD thesis, University of Port Elizabeth

Coppejans, E., Richmond, M.D., De Clerck, O. & Rabesandratana, R. (1997). Marine macroalgae. Seaweeds. In: *A Guide to the Seashores of Eastern Africa and the Western Indian Ocean Islands*, ed. M.D. Richmond, pp. 70–95. SIDA, Department for Research Cooperation, SAREC, Stockholm

Dahdouh-Guebas, F. & Koedam, N. (2001). Are the northernmost mangroves of West Africa viable? — a case study in Banc d'Arguin National Park, Mauritania. *Hydrobiologia* 458: 241–253

Grandvaux Barbosa, L.A. (1970). *Carta fitogeographica de Angola.* Luanda, Inst. Invest. Ciet. Angola.

Hogarth, J. (1999). *The biology of mangroves.* Oxford University Press

Hughes, R.H. & Hughes, J.S. (1992). *A Directory of African Wetlands.* IUCN, Gland Switzerland and Cambridge UK/ UNEP, Nairobi, Kenya/ WCMC, Cambridge, UK

Jaasund, E. (1976). *Intertidal Seaweeds in Tanzania, a Field Guide.* Univ. Tromsø, Tromsø

Jenik, J. (1970). Root systems of tropical trees 5. *Preslia Praha* 42: 105–113

Kairo, J.G., Dahdouh-Guebas, F., Gwada, P.O., Ochieng, C. & Koedam, N. (2002). Regeneration status of mangroves forests in Mida Creek, Kenya: a compromised or secured future. *AMBIO* 31 (7–8): 562–568

Kalk, M. (1995). *A Natural History of Inhaca Island Mozambique.* Witwatersrand Univ. Press, Johannesburg

Kokwaro, J.O. (1985). The distribution and economic importance of the mangrove forests of Kenya. *J. East Afr. Nat. Hist. Soc. Nat. Museum* 75(188): 1–10

Lawson, G.W. (1985). Algae associated with mangroves in the Niger delta area. In: *The Mangrove Ecosystem of the Niger Delta.* ed. B.H.R. Wilcox & C.B. Powell, pp. 56–67. University of Port Harcourt

Lugomela, C. (2002). 'Cyanobacterial diversity and productivity in coastal areas of Zanzibar, Tanzania'. PhD thesis, Stockholm University

Macnae, W. & Kalk, M. (1962). The ecology of the mangrove swamps of Inhaca Island, Mozambique. *J. Ecol.* 50: 19–34

Moses, B.S. (1985). Mangrove swamp as potential food source. In: *The Mangrove Ecosystem of the Niger Delta*, ed. B.H.R. Wilcox & C.B. Powell, pp. 170–184. University of Port Harcourt

Odu, E.A. (1985). Preliminary studies on the bryophyte flora of the mangrove and fresh-water swamp forests in the Niger delta. In: *The Mangrove Ecosystem of the Niger Delta*, ed. B.H.R. Wilcox & C.B. Powell, pp. 68–87. University of Port Harcourt

Powell, C.B. (1985). The decapod crustaceans of the Niger delta. In: *The Mangrove Ecosystem of the Niger Delta*, ed. B.H.R. Wilcox & C.B. Powell, pp. 226–238. University of Port Harcourt

Roger, E. & Andrianasolo, M. (2003). Mangroves and salt marshes. In: *The Natural History of Madagascar*, ed. S.G. Goodman & J.P. Benstead, pp. 209–210. University of Chicago Press, Chicago

Semesi, A.K. (1993). Mangrove ecosystems of Tanzania. In: Conservation and sustainable utilization of mangrove forests in Latin America and Africa regions. Part II — Africa, ed. Diop, E.S. International Society for Mangrove Ecosystems and Coastal Marine Project of UNESCO. Mangrove Ecosystems Technical Reports volume 3

Semesi, A.K. (1998). Mangrove management and utilization in Eastern Africa. *AMBIO* 27: 620–626

Spalding, M.D., Blasco, F. & Field, C. (eds) (1997). *Mangroves Atlas of the World*. International Society for Mangrove Ecosystem, Okinawa, Japan

Sunderland, T.C.H. & Morakinyo, T. (2002). *Nypa fruticans*, a weed in West Africa. *Palms* 46 (3): 154–155

Taylor, M., Ravilious C. & Green, E.P. (2003). *Mangroves of East Africa*. UNEP WCMC, Cambridge.

Tomlinson, P.B. (1986). *The Botany of Mangroves*. Cambridge University Press

Vannini, M. & Ruwa, R.K. (1994). Vertical migration in the tree crab *Sesarma leptosoma* (Decapoda, Grapsidae). *Marine Biology* 118: 271–278

Walter, H. & Steiner, M. (1936). Der Okologie der ostafrikanischen Mangroven. *Z. Bot.* 30: 65–193

White, F. (1983). *The vegetation of Africa*. UNESCO, Paris.

Photo credits

Salomão Bandeira: pages 8, 9 (upper), 17, 18, 19, 20, 45, 54, 66, 67, 70, 71, 84

Laurence Dorr: pages 10, 21, 74, 75

Andrew McRobb: front cover, pages 6, 9 (lower), 40

Martin Cheek: pages 55, 80

Index to local names

Abuma	*Rhizophora racemosa*
Afiafy	*Avicennia marina*
Aguirigui	*Avicennia germinans*
Akocut	*Rhizophora racemosa*
Ambianbiolan	*Laguncularia racemosa*
Antsombera	*Barringtonia asiatica*
Aysilola	*Bruguiera gymnorhiza*
Bak	*Laguncularia racemosa*
Bandjo	*Avicennia germinans*
Black Mangrove	*Avicennia germinans*
Bois d'amande	*Pemphis acidula*
Bue	*Avicennia germinans*
Bue-Dinte	*Avicennia germinans*
Cava	*Ceriops tagal*
Chemchem-de	*Laguncularia racemosa*
Dengi	*Rhizophora harrisonii*
Diain mangui	*Laguncularia racemosa*
Egba	*Rhizophora racemosa*
Fanavitraina	*Avicennia marina*
Farafaka	*Ceriops tagal, Sonneratia alba*
Farahafaka	*Sonneratia alba*
Fotabe	*Barringtonia asiatica, Barringtonia racemosa*
Fotadrano	*Barringtonia racemosa*
Fotaka	*Barringtonia asiatica*
Fotatra	*Barringtonia racemosa*
Gama	*Pemphis acidula*
Gandel	*Rhizophora mucronata*
Gbeleti	*Avicennia germinans*
Ginni Kandala	*Ceriops tagal*
Guitouconon	*Avicennia germinans*
Hafihafy	*Avicennia marina*
Hlohlodjani	*Ceriops tagal*
Hlothlotxwani	*Ceriops tagal*
Honkofotsy	*Avicennia marina*
Honkolahy	*Avicennia marina, Ceriops tagal, Rhizophora mucronata*
Honkovavy	*Rhizophora mucronata, Ceriops tagal*
Iguiri	*Avicennia germinans*
Ikapa	*Bruguiera gymnorhiza*
Ikapa	*Ceriops tagal*
Ikapa	*Rhizophora mucronata*
Inetanda	*Rhizophora racemosa*
Inetande	*Rhizophora racemosa*
Invede	*Avicennia marina*
Jaia-guwi	*Avicennia germinans*
Jaia-Lelei	*Rhizophora harrisonii*
Jaiei	*Rhizophora harrisonii*
Ka bure	*Avicennia germinans*
Kandalo	*Rhizophora mucronata*

Kapa	*Ceriops tagal*
Kikandaa	*Lumnitzera racemosa*
Kilalamba kike	*Pemphis acidula*
Kinsi	*Rhizophora harrisonii*
Kinsii	*Rhizophora racemosa*
Kinsu	*Rhizophora racemosa*
Lavasiko	*Bruguiera gymnorhiza*
Lichochochana	*Ceriops tagal*
Lovingo	*Lumnitzera racemosa*
Madaroboka	*Barringtonia racemosa*
Maganga	*Laguncularia racemosa*
Mahondro	*Barringtonia racemosa*
Mangal branco	*Avicennia marina*
Mangal indiano	*Ceriops tagal*
Mangal maça	*Sonneratia alba*
Mangal preto	*Lumnitzera racemosa*
Mangal vermelho	*Rhizophora mucronata*
Mangal-bola-de-canhão	*Xylocarpus granatum*
Mangal-Mozambique	*Heritiera litoralis*
Mangaoka	*Barringtonia racemosa*
Mangui	*Rhizophora racemosa*
Manko	*Rhizophora harrisonii, Rhizophora mangle*
Manondra	*Barringtonia racemosa*
Manondro	*Barringtonia racemosa*
Marrubo	*Xylocarpus granatum*
Masotry	*Avicennia marina*
Mbougari	*Avicennia germinans*
M'Candalla	*Rhizophora mucronata*
Mchonga	*Bruguiera gymnorhiza*
Mchu	*Avicennia marina*
Megama	*Pemphis acidula*
Meshuz	*Avicennia marina*
Metinindi	*Sonneratia alba*
M'fava	*Xylocarpus granatum*
M'Fingi	*Rhizophora mucronata*
M'Finji	*Bruguiera gymnorhiza*
M'finse	*Bruguiera gymnorhiza*
M'fumansi	*Bruguiera gymnorhiza*
M'gama	*Pemphis acidula*
Mgando	*Rhizophora mucronata*
Mianda	*Ceriops tagal*
Miarco	*Avicennia germinans*
Mkaa pwani	*Pemphis acidula*
Mkamkam	*Pemphis acidula*
Mkandaa	*Ceriops tagal, Lumnitzera racemosa*
Mkandaa mwekendu	*Ceriops tagal*
Mkandaa mwitu	*Lumnitzera racemosa*
Mkandaa ya pwani	*Ceriops tagal*
Mkandaa-dume	*Lumnitzera racemosa*
Mkapa	*Sonneratia alba*
Mkaracha ya pwani	*Ceriops tagal, Lumnitzera racemosa*

Mkoko mpia	*Sonneratia alba*
Mkoko	*Rhizophora mucronata*
Mkomafi	*Xylocarpus granatum*
Mkomasi	*Xylocarpus granatum*
Mliana	*Rhizophora mucronata*
Mlilana	*Sonneratia alba*
Mnyinyuwa	*Pemphis acidula*
Moromony	*Heritiera litoralis*
Mosotro	*Avicennia marina*
Mpedge	*Avicennia marina*
Mpira	*Sonneratia alba*
Mpiria	*Bruguiera gymnorhiza,*
	Sonneratia alba
Mpiripito	*Lumnitzera racemosa*
M'poronda	*Bruguiera gymnorhiza*
Msikundazi	*Heritiera litoralis*
Msindi	*Rhizophora mucronata*
Msinzi	*Bruguiera gymnorhiza,*
	Rhizophora mucronata
Msisi	*Bruguiera gymnorhiza*
Mtanganda	*Rhizophora mucronata*
Mtomondo	*Barringtonia racemosa*
Mtonga	*Xylocarpus granatum*
Mtongakima	*Rhizophora mucronata*
Mtoro-toro	*Barringtonia racemosa*
Mtovo-tovo	*Barringtonia racemosa*
Mtswi	*Avicennia marina*
Mtu	*Avicennia marina*
Mtulo	*Rhizophora mucronata*
Mucandala	*Ceriops tagal*
Muchwi mume	*Avicennia marina*
Muchwii	*Avicennia marina*
Mucoco	*Rhizophora mucronata*
Mucolongo	*Heritiera litoralis*
Mucomafi	*Xylocarpus granatum*
Muea	*Bruguiera gymnorhiza*
Mui	*Bruguiera gymnorhiza*
Muia	*Bruguiera gymnorhiza*
Murrubo	*Xylocarpus granatum*
Musisi	*Pemphis acidula*
Musso	*Avicennia marina*
Muvucuta	*Pemphis acidula*
Mwanawea mpwa	*Pemphis acidula*
Necolongo	*Heritiera litoralis*
Nhakandala	*Ceriops tagal*
Nhamesso	*Avicennia marina*
Nhantanzira	*Rhizophora mucronata*
Nipa	*Nypa fruticans*
N'kandaia	*Bruguiera gymnorhiza*
Nkandala	*Bruguiera gymnorhiza,*
	Ceriops tagal
Nsangi	*Ceriops tagal*
Nseti	*Xylocarpus granatum*
Ntame	*Rhizophora racemosa*
Ntehigizigi	*Laguncularia racemosa*
N'tsowozi	*Avicennia marina*
Nvandi	*Avicennia germinans*

Oellha	*Laguncularia racemosa*
Orke	*Laguncularia racemosa*
Oto	*Rhizophora racemosa*
Piripito	*Lumnitzera racemosa*
Red mangrove	*Rhizophora mangle*
Rongo	*Lumnitzera racemosa*
Saanar	*Avicennia germinans*
Salgheiro	*Avicennia marina*
Sanaj	*Avicennia germinans*
Sciavro	*Avicennia marina*
Sciori Guduo	*Rhizophora mucronata*
Sciori Medu	*Ceriops tagal*
Seane	*Xylocarpus granatum*
Séné	*Rhizophora racemosa*
Setaka	*Bruguiera gymnorhiza*
Shawerab	*Avicennia marina*
Shinkanga	*Rhizophora mucronata*
Shokuliha	*Xylocarpus granatum*
Shora	*Avicennia marina*
Showarab	*Avicennia marina*
Shule	*Rhizophora harrisonii*
Shundinte-le	*Rhizophora harrisonii*
Sinhaka	*Rhizophora mucronata*
Skasut	*Avicennia germinans*
Songery	*Sonneratia alba*
Sosuman	*Lumnitzera racemosa*
Sukire	*Pemphis acidula*
Sule	*Rhizophora harrisonii*
Suthe-le	*Rhizophora harrisonii*
Tanga foly	*Bruguiera gymnorhiza*
Tarafe preto	*Laguncularia racemosa*
Tarafe	*Rhizophora racemosa*
Thowozi	*Avicennia marina*
Tjindiri	*Sonneratia alba*
Tsingafiafy	*Avicennia marina*
Tsitolona	*Bruguiera gymnorhiza*
Tsitolonina	*Bruguiera gymnorhiza*
Txomahati	*Avicennia marina*
Unthoboti	*Avicennia marina*
Utovoji	*Avicennia marina*
Vahona	*Bruguiera gymnorhiza*
Vahona	*Lumnitzera racemosa*
Vahona	*Sonneratia alba*
Vonjioko	*Pemphis acidula*
Votishonko	*Lumnitzera racemosa*
White mangrove	*Laguncularia racemosa*
Xinkaha	*Rhizophora mucronata*
Xitaka	*Bruguiera gymnorhiza*

Index to scientific plant names